# THE
# MASTER
# OF
# CONFESSIONS

## ALSO BY THIERRY CRUVELLIER

*Court of Remorse:*
*Inside the International Criminal Tribunal for Rwanda*

# MASTER OF CONFESSIONS

## THE MAKING OF A KHMER ROUGE TORTURER

# THIERRY CRUVELLIER

## TRANSLATED FROM THE FRENCH BY ALEX GILLY
FROM *LE MAÎTRE DES AVEUX*
PUBLISHED IN FRANCE BY EDITIONS GALLIMARD

AN IMPRINT OF HARPERCOLLINS*PUBLISHERS*

LE MAÎTRE DES AVEUX © 2011 by Thierry Cruvellier. Originally published in France by Editions Gallimard/Versilio.

THE MASTER OF CONFESSIONS. Copyright © 2014 by Thierry Cruvellier. Translation copyright © 2014 by Susanna Lea Associates. All rights reserved. Printed in the United States of America. No part of this book may be used or reproduced in any manner whatsoever without written permission except in the case of brief quotations embodied in critical articles and reviews. For information address HarperCollins Publishers, 10 East 53rd Street, New York, NY 10022.

HarperCollins books may be purchased for educational, business, or sales promotional use. For information please e-mail the Special Markets Department at SPsales@harpercollins.com.

FIRST EDITION

Library of Congress Cataloging-in-Publication Data has been applied for.

ISBN 978-0-06-232954-7

14 15 16 17 18   OV/RRD   10 9 8 7 6 5 4 3 2 1

# CONTENTS

## NOTE TO THE READER

All citations and testimonies are quoted from the trial unless otherwise noted. Testimonies have been edited and condensed for clarity.

Please refer to page 319 for a historical overview of Cambodia and to page 325 for a detailed note on sources.

# CHAPTER 1

*My name is Kaing Guek Eav. I took the name Duch when I joined the Revolution. I wanted to liberate my people—my parents, my family, myself. Instead, my country was engulfed by tragedy and more than 1.7 million people died. As a man—as someone who believes in justice—I see now that the party I belonged to, the Communist Party of Kampuchea, was responsible. But back then, you couldn't challenge it. There was no way out. I had to follow orders.*

*My main objective was to interrogate people. I never killed anyone with my own hands. If I hadn't been there, someone else would have taken my place. But it was me: I had a pen, I made notes, I tried to write impartial reports to submit to my superiors, but they wanted confessions that led to more arrests. I sacrificed everything for the Revolution, and back then I believed in what I was doing. I was proud at the time. But as I look back now, it makes me shudder. The fact that I killed more than twelve thousand people makes me feel ashamed.*

Like many Khmers, Duch has a small build. His narrow shoulders make him seem especially slight. He's sixty-seven years old and, like many old men, has a potbelly, which he covers by hiking up his pants to his navel, stretching them up and over his paunch rather than buckling his belt beneath it. He moves quickly but with a stiffness in his chest and arms, which is perhaps the result of his years as a soldier of Communism, or perhaps simply from the passage of time.

The courtroom has a large, wooden, horseshoe-shaped witness stand. When speaking from it, Duch tends to focus on some indeterminate spot up to his left, which the filmmaker Rithy Panh calls his "blind spot." Panh tells me he does this because it helps him concentrate and stay in control.

Duch tends to breathe heavily through his nose mid-sentence, giving the impression that he's speaking underwater or through an oxygen mask. He's not known to suffer from asthma or any other respiratory trouble, yet sometimes after speaking he'll falter with his mouth open, as though trying to catch his breath. When he is nervous—which rarely happens—he rubs his face vigorously with the palm of his hand.

> *When I joined the Revolution, I was trained for secrecy. You had to keep your supervisor's identity secret; you didn't reveal who your subordinate was. If you killed people, or if you ordered people killed, that had to remain a secret, along with the number of people. Later, I heard the saying: "The better you keep secrets, the better you survive." We used to say that half the battle was keeping things secret. Say nothing, hear nothing, see nothing. Secrecy: that was my top priority when I was training people. The recruits were no good—they talked a lot. But I was very strict. Even my deputy was no match for me.*

Duch sucks air through his nose.

"That's it."

Age has weathered Duch's face favorably. His ears seem too large for his head and peel away from it at their tips, emphasizing the sharp angles of his face. The dark tint of his full and well-defined lips looks almost purple on the television screen. Sometimes, when he's sitting there with his mouth hanging open, waiting for the interpreter to finish, his eyes narrow until his pupils disappear and his high cheekbones accentuate the bags beneath them, and true age marks his face. His smile reveals crooked and severely decayed teeth and lends him a youthfulness that, in Duch's case at least, is less flattering than old age. His gaze is intense but strangely veiled, bright and glassy at the same time. Arched in the middle, as though by invisible pins, his discreet eyebrows form an inverted V, giving his eyes a haunted look, as in a person who has recently had cataract surgery. On any given day, the lines on his face can appear either deeply engraved or ironed flat, his eyes either wide-open or shrunken to slits. Once, while trying to identify the wife of a witness, he puckered his mouth until his lips curled at the corners, and I couldn't tell if he was expressing skepticism or stifled contempt. Both reactions come naturally to him.

Though generally stiff and erect, Duch does have his offhand moments. At one point, while listening to a former subordinate, he smiles, folds his foot beneath him, and slumps into his seat, as though abruptly released from his usual sense of decorum. From the witness box, the former guard describes Duch as "a firm, serious, and very meticulous man," a heavy smoker with whom the guard had never dared joke around but whom he had seen erupt into fits of laughter at least once.

"Are you frightened of him now?" asks one of the five judges.

"No. I'm not frightened of him."

When another one of the judges, a woman, describes Duch as intelligent, educated, hardworking, enthusiastic, attentive to detail, methodical, professional, eager to excel, willing to please his superiors, and generally proud of his work, Duch agrees.

Duch is also endowed with a prodigious, though selective, memory, as well as a bitter sense of history.

He notes, "The words 'meticulous,' 'hardworking,' and 'determined' used to describe me would be considered virtues if spoken in the context of a government that loves its country and its people. But the government I served was the opposite: it was a cruel and criminal machine, and in this context, these words are painful to hear."

It only takes Duch a moment to gauge his interlocutor and adjust his behavior accordingly. To a respectful young Khmer woman asking challenging questions, he lowers his voice almost to a murmur, whereas to a hostile foreign lawyer he replies in a cold and confident tone. He knows how to adapt his speech and rhetoric. In a single session, for instance, he delivers long and convoluted responses to a

Cambodian prosecutor, then gives a brief aside to a foreign prosecutor, then nothing but curt and sharp answers to a European lawyer representing civil parties.

Duch is generally prepared to cooperate and provide details. But he's also perfectly capable of giving terse yes or no answers, which the prosecutors and victims' representatives try unsuccessfully and often ineptly to draw out. For almost eight years, Duch's job was to interrogate his adversaries: to make them talk, either voluntarily or by force. His grasp of human psychology, of the dynamics of interrogation and power relationships make him a tough, well-armed opponent at his own trial. And he can count on Kar Savuth and François Roux, the Cambodian and French lawyers, respectively, who constitute his experienced and coordinated legal team.

If his interrogators in the courtroom prove diffident or incompetent, Duch quickly takes the upper hand. And even if he doesn't win the battle, his opponent nevertheless loses it, which brings a scornful smirk to Duch's face. Appearing confident and relaxed, Duch is dismissive of the counsel for civil parties, whom for the most part he judges to be beneath him and who, smug because they have power and he does not, quickly find themselves lambasted, reduced, and demeaned by the very man they want to cut down. Duch will fall, of course—but when?

Duch sometimes gets carried away. For instance, when faced with a weak or ineffective line of questioning, he can react both snidely and with insolence—two highly unfavorable traits in a court of law. On rare occasions, he also feigns a clumsy false cheer, as when he affects a friendly attitude toward one of the three survivors from his prison. You feel embarrassed for him when, at such moments, he is overcome by nervous laughter and has to hold his hand over his mouth until it stops.

At times Duch is also impressionable. In a rare moment when the trial's focus veers from its usual litany of barbarism to a disputed point of actual law, the lawyers and prosecutors are visibly delighted. Glee illuminates the faces of the more eloquent among them as they indulge in a little courtroom grandstanding. When one of them leans down to whisper to a colleague, he brings to mind a giant black flamingo, his robes flowing about him, elbows thrust back like folded wings. Duch, sitting silently behind his legal team, clearly enjoys and admires the legal sparring and nimble mind games.

The former keeper of the Party's secrets is as energetic and talkative as he is loath to display emotion. Still, some circumstances and names get to him. When they do, you hear him swallow and sniff, you see his jaw clench and his lower lip suck the upper one in, you hear a muffled groan and see his face contort as he fights back the tears. He stays like that, his upper lip shrouded by the lower one as though stuck to his teeth, his eyebrows raised, his eyes wide-open and pleading for help. On one Friday, after a week of particularly gruesome testimony, Duch again speaks of being crushed by shame. He turns away with his eyes to the ceiling, and you can see the turmoil he's in.

"I'm stopping now," he says without breaking down.

Yet though his shield has been pierced and it seems he has at last been broken, Duch proves remarkably resilient—he shows up the following Monday, looking not just strong but defiant.

TWO IMAGES OF DUCH stand out in my memory. The first is of the moment when, while recounting to the court the day he swore total loyalty and devotion to the Communist Party of Kampuchea (the Khmer name for Cambodia), he stood and gave the official revolutionary salute, his arm bent at a right angle, his closed fist held level with his head, with an intensity and conviction that appeared undiminished some thirty years after the fall of the regime. It's a terrifying image, one that reveals the depth of conviction possessed by a man described in court by psychologists as capable of entertaining "only one idea and only one thought at a time."

The other image was captured before Duch's trial began. During the pretrial investigation, the investigators wanted to interrogate Duch in the S-21 prison, the death mill he managed in Phnom Penh from 1975 to 1979. Over the course of a long, painful, and laborious morning, Duch, the three still-living survivors of S-21, and a handful of former guards, interrogators, and torturers tried to "reconstruct" the crime scene. The foreigners working alongside the Cambodians at the court were already sweating in the February sun, though it wasn't yet the furnace that is Phnom Penh in April, when light vaporizes the city's colors into one sultry haze. It was approaching noon, and Duch was standing in the middle of one of the interior courtyards at S-21. His brow was low and straight, without the furrow that gives him a haunted look. In his eyes, you could detect torment caused by some painful question. His eyelids crinkled so that they looked like small waves, gentle rollers washing ashore at his temples. His half-open mouth allowed a glimpse of his unattractive teeth. His face had an unresolved rather than tense expression. It was then that the hard-line Communist official disappeared and an old man ravaged by inner demons appeared in his place. Duch looked up at the sky, torn between the fear of punishment and the urge to cry.

Back in court, facing his judges and the public and supported by his calm, steadfast legal team, Duch asks:

> My intention was to transform from an ordinary man into a Communist man. It was 1964. I became a new man called Duch, different from the math teacher called Kaing Guek Eav. Today, I declare before the world my intention to change back into an ordinary man. Now that we are in the midst of this trial, facing this tribunal, do you see me as a new man?

# CHAPTER 2

**T**WO OR THREE MONTHS AFTER KAING GUEK EAV WAS BORN IN November 1942, a fortune-teller told his parents he was worried about the child's name. It didn't augur well, he said. It made the child vulnerable to illness. So, pressured by this prediction, Kaing Guek Eav's parents changed his name to Yim Cheav. But by the time the child reached his teens, he disliked both his new name and the fortune-teller who was responsible. For him, the name signified being "slow, poor, outdated, a straggler." At the age of fifteen, he asked two things of his father: first, to allow him to take back his original name; and second, to change his date of birth: he had started his studies late, at the age of nine, and he wanted to seem younger in order to be able to take his exams.

Changing one's name or date of birth is common practice in Cambodia, where no one celebrates birthdays anyway. Time doesn't accrue here, it cycles: if it just goes around in a circle, there's no point keeping count.

Ten or so years later, Kaing Guek Eav again felt compelled to change his identity. This time, he wanted to become a Communist. He wanted to be a *new man*:

> *My name was Chinese and I needed a Khmer name. I chose the name "Duch" because I liked it. There was a sculpture of Buddha carved by a great sculptor called Duch, whom my grandfather held in high esteem. So I drew inspiration from his name. In the first text I ever read at primary school, Duch was also a diligent student, very*

*obedient and praised by the teacher. That's why I liked the name Duch.*
*It belonged to someone good and was also a Khmer name.*

Duch, in Khmer, isn't pronounced "dutch." "Uch" is an arcane phonetic spelling that is supposed to sound like an open "oïk," as though there's a catch at the end. "Duch," in other words, is the Khmer way of writing what is pronounced "doïk," just as you don't say "Khmer" but "kmay." Linguists have developed a way of writing sounds that is comprehensible only to the initiated. They aren't the only ones trying to protect their knowledge from the common man. Lawyers also have their own phonetics: the law. The legal profession likes to pronounce something just or lawful in terms that make the pronouncement clear only to those in the profession. Insiders—whether in linguistics, law, or politics—are wary of the autonomy of their fellow man.

So we read as "Duch" a name that we pronounce "Doyk." Paradoxically, however, the linguists have given everyone the freedom to address Duch as they see fit. Thus, the French judge calls him "Dook" while the judge from New Zealand addresses him as "Mr. Kaing Gek Yu," the correct pronunciation of the phonetic (and therefore misleading) "Kaing Guek Eav." One prosecutor says "Mr. Kaing" while another lawyer says "Mr. Dook."

Revolutionaries often have multiple identities. What is uncommon is for one to repent of his crimes. In less than four years in the late 1970s, the Communist Party of Kampuchea annihilated between a quarter and a third of the population of Cambodia. Yet the "brothers" who ran the Party have always insisted that they had nothing to do with the massacre. Duch, who was their direct subordinate, is the only high-ranking Khmer Rouge cadre to have admitted his part in the destruction of his people.

Duch's admission of responsibility is "pretty close to unique among surviving active members of that administration," historian David Chandler tells the tribunal.

That Duch admits both to the bulk of his crimes at S-21 and to the criminal nature of the ideology he served makes his trial unique;

that he stood up in court every day for six months to explain himself and his actions makes it even more so. Not a day passed when the defendant did not address the court, and no question of fact or point of history was examined without being put to him. I have covered several trials for genocide or crimes against humanity in international courts; no other perpetrator has been given such ample opportunity to be heard—not in Arusha, Freetown, or The Hague.

In Phnom Penh, Duch was the only one to behave like this. At the time of his trial, four other Khmer Rouge leaders were to stand before the same tribunal after him. All of them were in their eighties. All of them denied everything. By December 2013, only two of the four accused were still facing trial, as one was declared mentally unfit and another had died.

**THE TRIBUNAL TASKED WITH** trying the Nazis was set up, naturally enough, in the Nuremberg Palace of Justice. In Arusha, the court deciding the case against the perpetrators of the Rwandan genocide chose as its seat the conference center where peace talks had taken place. In The Hague, the tribunal for Lebanon is in a building formerly occupied by the intelligence service, while the International Criminal Court will soon take up residence in what were once military barracks. Each of these tribunals chose as its seat a location that is more or less symbolically apt.

In Phnom Penh, the Khmer Rouge Tribunal changed venue at the last minute. The trials were originally supposed to have been held in the historic Chaktomuk Conference Center in the center of town. Built in the early '60s by the master of New Khmer Architecture, this great hall is located on the banks of the Tonle Sap River. Its eight-point serrated roof makes it look like a giant, handheld fan, or perhaps a palm frond, while the spire soaring from the vertex of its triangular shape brings to mind a giant compass. It was here, shortly after the fall of the Khmer Rouge in 1979, that Pol Pot was tried in absentia, in a trial organized under the Vietnamese occupation and subject to the vagaries of the propaganda of the time. It was here, too, that twenty-four years later, the United Nations and the government of Cambodia signed an agreement to create a tribunal to try the handful of surviving high-ranking Khmer Rouge leaders.

Including Duch.

But at the last minute, the government decided that noble Chaktomuk Hall wasn't spacious enough and that holding the trials there would cause traffic problems. So the government generously suggested—or rather decided—to move the tribunal to a military base on the outskirts of town, some forty minutes by car from the city center. In symbolic terms, there's something almost wanton about the turnaround.

However this exile from the city center has done the tribunal no harm in terms of space or attendance. Its public gallery is by far the largest and most comfortable of the seven international courts established in Africa and Europe over the past two decades. In fact, the five-hundred-seat amphitheater is so vast that we observers end up watching much of the proceedings on the flat-screen televisions installed in the gallery, rather than directly. Witnesses in the courtroom have their backs to us when they take the stand, so we only see their faces on the screens. It may seem strange, but we watch on television the trial taking place in the courtroom before us.

Every day, dozens of flashlights, plastic water bottles, pots of Tiger Balm, cigarette lighters, and various other provisions accumulate on the shelves next to the metal detector at the entrance to the public gallery. Hundreds of villagers are bused in by the tribunal's Public Affairs Section or by local associations. One of the first things that these villagers learn when coming face to face with international justice is that international justice considers dangerous or discourteous items that are practical or essential for villagers: water, ointment, and newspapers are not allowed.

Three flags hang above the judges' heads: that of the Kingdom of Cambodia, with its restrictive motto "Nation, Religion, King"; that of the United Nations, with its fragile olive branches of peace;

and that of the tribunal itself, with its cumbersome name—the Extraordinary Chambers in the Courts of Cambodia—on which the UN olive branches curl around a Khmer prince from Angkor times sitting cross-legged and holding a sword in his right hand, tip pointed to the sky. The judges, three Cambodians and two Westerners, thus find themselves under allegiance to three discrete entities: to their country (or host country), to the United Nations, and to themselves. Some say that holding multiple allegiances keeps a person from making extremist choices. This precarious triple fealty, however, hovers over the judges like damnation over the heads of churchgoers.

To enter the gallery, spectators must pass through two metal detectors. Once inside, a massive, soundproofed, plate-glass window separates them from the courtroom. Five guards stand sentry inside the vast public gallery.

If repression can be ranked by degrees, then the tribunal's security detail is certainly at the lower, more benign end of the scale. The sentries in Phnom Penh are nothing like those at the International Criminal Court in The Hague, who, elsewhere and in other circumstances, wouldn't seem out of place in the darkest of militias. As for those guarding the tribunals of the Third World, a cheerful nonchalance often belies their uniforms and regulations. A Dutch guard is much ruder and infinitely more hostile than a Khmer, Sierra Leonean, or Tanzanian one. Wherever they're from, though, they're all exposed to the same crushing boredom.

There is little chance of any trouble arising at the tribunal, and none at all of an attack. But if there's no threat of trouble, then it must be prevented with even greater zeal. The tedium is as great for the public as it is for the tribunal staff, and one way to break it is to ban something new. One nuisance specific to Phnom Penh is the ban on Tiger Balm, an ointment as precious to those who work in Khmer fields as lipstick is to Parisian women. Yet a guard at the last checkpoint before the courtroom quickly ferrets out the aromatic rub.

Inside, some guards work just as zealously to impose a proper sense of decorum on the public. Shutting your eyes is forbidden, as is raising a knee above the back of the seat in front. Letting your eye-lids droop was also banned at Nuremberg in 1945, as was crossing your legs if you were sitting in the front row. Yet despite the intense security during the Nazi trials, the journalist Rebecca West described how one of her female colleagues once smuggled into the courtroom a loaded pistol in her jacket sleeve. Nothing so sensational happened in Phnom Penh. But the occasional buzz of a vibrating cell phone, or whiff of menthol, or magazine sticking out from beneath a notebook reminds us, with reassuring regularity, of how one can always make a mockery of law and order.

Most of the people in the public gallery have skin the color of mahogany, of burnt umber or old leather—colors that give them away as country folk.

Their presence alone is a blithe challenge to the endless crush of rules and regulations. The only thing that equals the surprise a Khmer peasant feels when his Tiger Balm is confiscated is his bewilderment on being scolded for napping. One day a woman falls asleep on her neighbor's shoulder. She can't help it: she had to leave her village at one in the morning to make the session. The guard tries to shake her awake. He fails and, flummoxed, gives up. Old farming women, as supple as they are slight, curl up on their chairs in that position so natural to Khmers but so awkward for everyone else: with their legs folded back, in parallel and off to one side so as not to offend Bud-dha. And not even the most zealous guard dares prevent people from kicking off their sandals. In Asia, even the rich go barefoot.

For those bused in from the nation's rice paddies, no courtroom rule stays sacred for very long. With a blissful lack of awareness, they ignore the rule about not standing until after the president of the tribunal has stood, just as they ignore the diktat that no one should leave until the last judge has exited the court. From the first recess on the first day of the trial, the guards are spectacularly overwhelmed, and there's a cheerful, gratifying buzz when, much to the guards' con-sternation, everybody gets up at once. It's a metaphorical victory of the people over the mighty and a refreshing sight, like a revolution without the dogma, or a massive jailbreak. Throughout the rest of the trial, the guards never once succeed in calming the ruckus kicked up by these common folk. Watching the guards, arms dangling by their sides, stumped by their inability to corral the cheerful flood of people, is a daily and secret pleasure; one that lets you believe, even for a fleeting moment, in freedom.

One day, while Duch is giving a painstaking analysis of the Party's propaganda machine in the courtroom, an eye-catching group of observers swarms into the public gallery. All of them are wearing the same T-shirt emblazoned with the name of the tribunal, and their presence makes the gallery feel like a more cheerful rendition of a Communist party meeting. Scores of baseball caps, T-shirts, and notebooks bearing the court's name had just been manufactured and distributed. Present-day Cambodia is run by former Communists, including some notorious erstwhile Khmer Rouge, and certain habits, such as producing propaganda for the masses, die hard. The four hundred people brought in every day from different parts of the country or from the schools and universities of Phnom Penh sometimes seem like a perfect example of mass political mobilization.

Still, despite all its quirks, Duch's trial will give thirty thousand Cambodians the chance to spend at least a day inside this court, unique in their country. No other international trial has had an audience as vast and wide-ranging as this one.

**THE CRIMES COMMITTED BY** the Khmer Rouge are thirty-five years old, and the trial draws members of at least three generations. First, there are those who came of age in the 1970s, when the Communist guerrillas seized power. For the Cambodians among them, this was their greatest misfortune. For many Western Communist sympathizers of that generation, the rise of the Khmer Maoists in the midst of the Cold War became a focal point of their political activism—until it transformed their utopia into a killing field.

Then there are those who came of age in the 1980s and for whom

Pol Pot, deposed but still a threat, stood alongside Stalin and Hitler to complete the twentieth century's blood-soaked totalitarian triumvirate.

Finally, there are those born while international Communism was dying its ugly death and who learned about Marxism-Leninism the way you might learn about steam engines, with their old-fashioned jargon. For them, the most interesting thing about the twentieth century's blood-stained ideological experiment is the case studies it now provides, where we can see international justice at work.

All these disparate elements converge around Duch's case. I was born the year Duch swore allegiance to the Communist Party. I was twelve years old when the Vietnamese Communists put an end to his crimes, twenty-two when the Berlin Wall fell, and thirty-one when Pol Pot died and Cambodia's civil war, then as old as me, ended. Many in the gallery had personal reasons to be here. I had none other than having turned twenty years old during the Cold War.

This trial brings us all together. Sometimes we connect, sometimes we avoid each other—but all of us are in it together.

# CHAPTER 3

**B**OU MENG WAS TWENTY-EIGHT YEARS OLD IN 1970, THE YEAR he answered Prince Norodom Sihanouk's rallying cry against the forces that had just deposed him. The following year, Bou Meng went into the *maquis*, bands of guerrilla fighters, by then controlled by the enigmatic Khmer Rouge. The fledgling revolutionary movement was quick to make use of his artistic talents, and he soon found himself painting portraits of Marx and Lenin, mimeographs of which were distributed to Khmer Rouge combat units so that their fighters could recognize the founding fathers of Communism. Four years later, on April 17, 1975, the Khmer Rouge entered Phnom Penh. Bou Meng cheered the victory, but his cheers turned to dismay when the movement forced the capital's entire population to evacuate. The following year, his superiors were arrested, and Bou Meng started losing confidence in this revolution that rewarded its soldiers so poorly.

"I wore the black shirt, but my spirit wasn't in it," he tells the court.

In the land of the Khmer Rouge, when a commander was arrested, his men soon followed. It was known as a "line." A few months after the fall of his commander, Bou Meng and his wife were transferred to what he dubs a "hot reeducation" cooperative: in effect, a forced labor camp run with ruthless discipline. Like hundreds of thousands of his countrymen, Bou Meng became a prisoner. He dug canals and built dykes until he was on the verge of collapse. Then he had the good fortune to be transferred first to carpentry, then to the vegetable garden. He grew cabbages and eggplants for the collective. In May of 1977 (or maybe it was June, he doesn't remember exactly) he was slaving away in a vegetable patch when a group of black-shirted men jumped out of a jeep like a murder of crows. They told Bou Meng and his wife to pack their things; they were going to become teachers at the School of Fine Arts. Bou Meng was thrilled—he was a painter, not a gardener. He and his wife cheerfully got into the vehicle. The vehicle drove away from the camp, then stopped. They were ordered to get out, to sit down, and to put their hands behind their backs. They were tied up and blindfolded. Bou Meng's wife began to cry. He sank to the depths of despair.

In the courtroom, Bou Meng pauses in his story. He brings a hand to his forehead, as though the ghosts of the past are pounding too hard, as though he's about to lose consciousness. Duch is in the dock, sitting upright and perfectly still.

Unlike Bou Meng, Vann Nath—also a painter—didn't serve in the army. He was just nineteen when the Khmer Rouge won the war. But on December 30, 1977, like Bou Meng, he was arrested by men in black by order of the Angkar—"The Organization," in Khmer—the secretive, all-knowing, and all-powerful body that controlled everything in the new "Democratic Kampuchea."

Vann Nath is just sixty-three when he takes the witness stand, but he looks feeble and tired. He's a tall man and he wears a billowing, pale yellow shirt. He greets the judges, the prosecutors, and the defense. Duch doesn't move. The painter's hair is cut very short and has gone gray. His eyebrows, slightly disheveled at their outer edges but still black near his nose, are the predominant feature of his face, the roundness of which is emphasized by his full cheeks that have only just begun to sag with age. His deep voice contrasts with the presiding judge's high-pitched one. Vann Nath speaks with his eyes almost closed and glued to the ground. He continually massages his stomach. Even though he has told it countless times over the past thirty years, emotion overcomes Vann Nath almost as soon as he begins to tell his story. Like Bou Meng, he raises his hand to his forehead, grabs a handkerchief, and pulls himself together before continuing.

Vann Nath spent his first night of detention bound in leg irons in a pagoda-turned-prison. Then he was taken away on a motorcycle. Upon reaching his destination, he was interrogated for the first time. "You're a traitor," they told him. How many secret meetings had he held? "You'd better remember. The Angkar never makes mistakes." To help convince him, his interrogators pulled out electrical wires. Vann Nath saw bloodstains and plastic bags hanging on the wall. So, how many meetings? They gave him his first shock. He passed out. Someone threw water in his face. He came to. Then a second shock. He passed out again. Then another, and another after that. Afterward, he couldn't remember what answers he had given his torturers. He was ordered to get into a truck, where he was bound to six other men. There were eighteen prisoners in total. At around midnight, the truck pulled up somewhere. (They were on Street 360 in Phnom Penh, but Vann Nath didn't know that.) The prisoners were weak and exhausted. They couldn't stand. They were made to sit on the ground in two rows. Then they were roped together by their necks, blindfolded, and, despite their exhaustion, ordered to march single-file. Voices taunted them as they walked blindly, each man with his hand on the shoulder of the man in front. A question barreled around Vann Nath's head: what had he done wrong?

It was January 7, 1978. He had just entered S-21.


**IN THE CAMBODIAN CAPITAL,** emptied of its inhabitants, the secret police established a security perimeter around S-21 that extended far beyond the prison's single building. The prison itself constituted a small section of a much larger zone. A whole neighborhood was sealed off, with no one allowed in or out. Those who worked at S-21 lived, ate, and worked in the zone without ever leaving. Vehicles delivering prisoners didn't go straight to the gates of the detention center itself. They usually stopped somewhere in the vicinity in order to protect the absolute secrecy of its location.

Him Huy, a member of the guard unit, escorted newcomers from the arrival point to the prison itself, where he handed them over to Suor Thi, whose job it was to note down their "biography" and to register them in the system.

In the courtroom, Suor Thi looks like a bank teller. He sits straight, his face expressionless and smooth, his demeanor mechanical without being cold, as though he's put his smile away in a box labeled POINTLESS AND DAMAGING EXPRESSIONS. He sits with his arms crossed and his eyes lowered, perfectly still, showing no emotion. Only his constantly blinking eyelids break his otherwise statue-like stillness, though he does sometimes glance at the judge interrogating him:

> *After I took their names, the prisoners were sent to the photographers. Then they were blindfolded again and taken to the cells. I had to keep a record of which rooms in which building they were being held, so that we could keep track of the number of prisoners per cell and to make it easier for the interrogators.*

In the English interpretation, Suor Thi uses the word "rooms" to denote the shared cells. The prison clerk describes his workday in the same neutral, even tones that someone managing a large hotel might use to describe the number of short- and longer-term guests currently checked in. Once a photograph of a new prisoner had been taken, Suor Thi attached it to the short "biography" he had written. He was twenty-four years old, and his job was to keep the list of detainees at S-21 up to date, to record the names of the incoming and outgoing prisoners. In other words, he was the registrar of death.

Suor Thi didn't deal directly with the important prisoners, who were received separately. He was given their names for the register only later by Hor, the number two at the prison and the person in charge of its daily administration. Nor did foreign prisoners pass through Suor Thi's office. There was a different procedure when the personnel of S-21 themselves were arrested and thrown into irons in the very place where, the day before, they had been carrying out their tasks. They were led in with their faces covered so that their colleagues wouldn't recognize them. As for the children who ended up in S-21, there was no point writing down their biographies or taking snapshots of them, says Suor Thi: "I paid no attention to the children because I had to pay such close attention to the prisoners. None of the children would survive. All of them would be killed."

Suor Thi reminds the court that he was alone in his task and that sometimes his workload was considerable. He had to be available to the prison at all times, without exception. On a normal day, he would process between one and twenty people. But he remembers that in 1978, prisoners flooded in by the hundreds. It was during this period that some prisoners were photographed inside the cells themselves, an anomaly that was against the center's strict regulations. At the time, the engine of death was at full throttle, overheating even. In all the hustle and bustle, a new arrival might inadvertently be taken to their cell without having been photographed first. So they tried to work through the backlog by taking the photos inside the cells, explains the former registrar.

While Suor Thi is describing the registration and record-keeping processes, Duch is nodding. Leaning forward with his chest over the table and wearing an elegant white shirt, he shows no sign of scorn. There's a look of concentration on his face, of concern, a look he reserves for those he respects or deems legitimate. During the court's recess, he seems very relaxed, laughing with his Cambodian lawyer Kar Savuth, under the curious eyes of one of the two policemen guarding him.

**SUOR THI IS WEARING** a gray suit jacket that looks uncannily like the one worn by several witnesses over the past few days. The jacket looks well-made, but Suor Thi is drowning in it, just as those who testified before him were. A few days earlier, the tribunal employees in charge of preparing the witnesses needed a jacket. Pressed for time, they found the one Suor Thi is now wearing, forgotten by a prosecutor. They were happy with the result: not only did the jacket protect witnesses from the chill of the air-conditioning, but they looked nicer in it, too.

Like their compatriots in the public gallery, many of the witnesses are from the countryside. Their poorly made and badly cut shirts, the breast pockets misshapen by the objects they carry in them, give them away. Their shirts are sometimes so oversize you could fit two people in them, and their wrinkled collars fall limply. Through these shapeless clothes, you can make out undershirts clinging to their dry bodies.

At first, these simply dressed farmers brightened the sterile atmosphere that prevailed in the court. Then, midway through the trial, the court started making them wear the jacket, and the atmosphere lapsed back into that gray gloom where everyone looks the same, like detainees.

When a prisoner arrived at S-21 and had his photo taken, a number would be hung around his neck. Bou Meng's wife, Ma Yoeun, wore number 331, indicating that she was the 331st person to enter the prison that month. In her photo, Ma Yoeun wears her hair in the only style available to women in Democratic Kampuchea: a bob reaching halfway down her neck and parted square in the middle—though a few coquettish rebels pushed the part slightly to one side. In the photo, Ma Yoeun is a pretty twenty-five-year-old with a slightly frightened look in her eyes—a normal reaction in someone who has been arrested and just had her blindfold removed. The photo taken at S-21 is the only trace Bou Meng has left of his wife, and though it was taken by those who killed her, he never parts from it. She never leaves his wallet. Bou Meng has remarried, and his devoted second wife is twenty years his junior. But his memory is unfailing in its devotion to his first love.

After being photographed, Vann Nath and his fellow prisoners had their compulsory black clothes taken from them. The prisoners were left wearing nothing but their underpants. They were no longer worthy of the Glorious Revolution, and their clothes would be redistributed. This also prevented them from hanging themselves with the clothing.

Once the detainees had been identified, registered, photographed, and relieved of their revolutionary attire, Suor Thi handed them over to a guard, who took them to their cells. Sometimes he went to the cells himself to check prisoners off his list. But when he did, he says, he paid little attention to the conditions in which the men were incarcerated:

> *I know they suffered a lot. They were extremely thin, malnourished, and the air circulation was terrible. But I didn't really worry about it. My job was only to check them off the list and then hurry back to my office. I had just enough time to notice that they had become extremely weak.*

# CHAPTER 4

**T**HE S-21 PRISON WAS SET UP IN A FORMER HIGH SCHOOL. IT IS made up of five buildings shaped like a giant E. Buildings A, B, C, and D, which form the perimeter, rise over three floors, each with wide balconies running alongside the classrooms-turned-jail-cells. In the middle of the structure, the fifth block is a big, single-story house with a covered inner courtyard that divides the space into two distinct areas.

Bou Meng was incarcerated on the top floor of Building C. For months, he slept on the floor, weak with hunger to the point of feeling dizzy. When lizards crawled across the ceiling, he prayed they would fall on him so that he could eat them. Once, to his horror, the guards threatened to skin him alive. And like Vann Nath, like so many other prisoners in so many other prisons in Democratic Kampuchea, he kept asking himself, "What crime have I committed?"

Bou Meng shared the cell with about forty other detainees, including, for a while, a few foreigners. They all had long and dirty hair. They were covered in lice and infected sores. And though the guards tolerated no noise, they sometimes whispered among themselves. They were searched every night. Once a week—or maybe every fortnight, he can't remember exactly—they were hosed down. The floor flooded, so all the prisoners took off their shorts. Everyone was naked. Sometimes the guards would make fun of their genitals, remembers Bou Meng, apologizing to the court for mentioning it. They were treated worse than dogs or pigs, he says.

Vann Nath was thrown into a cell on the second floor of Building

B. In his recollection, the collective "shower" took place twice a week. Getting undressed while in leg irons was difficult. Those damned bindings were so uncomfortable, he remembers with a grimace. It took him thirty minutes just to carry out the maneuver. When a tactless judge asks him how he accomplished it, Vann Nath, as supple as ever, raises a leg to a right angle to demonstrate.

A month went by in these inhuman conditions. Sitting down without the warden's permission was forbidden. On a blackboard were chalked orders to not talk or to listen to the guards. The prisoners were served a meager bowl of gruel at eight in the morning and another at eight in the evening. They had to relieve themselves in the same room in which they were shackled, in an old munitions container, an iron box some fifteen centimeters deep. Vann Nath was covered with lesions. He couldn't stop scratching himself. He, too, hoped that a gecko would fall from the ceiling. But if it did, he'd have to gulp it down right away without being seen by the guard. If not, they would beat him. Unfortunately for him, Vann Nath was too far from the window, where the insects and lizards clustered: "Death loomed over us. People died one after another. They took the bodies away at ten o'clock. We didn't even care. We were like animals."

Vann Nath counted as many as sixty-five prisoners in his cell, lying on the ground in rows, their ankles shackled to a long metal rod. In one month, he saw four of his cellmates die. Sometimes the number in his cell fell to forty, when others were taken away and never seen again. The hardest thing was knowing that you hadn't done anything wrong, he says. The hardest thing was making up stories in order to survive, in order to avoid being tortured.

**"DID YOU CONSIDER IT** an ordinary type of job, or a special one?" Judge Lavergne asks a former prison guard on the witness stand.

"From what I saw, it was an ordinary type of job."

"What was 'ordinary' about it?"

"The Angkar assigned me to stand guard. I did the same thing all the time."

"If you were asked today to do an ordinary job of that nature, would you do it again?"

"No! I wouldn't!"

The public gallery bursts into laughter.

"What does the word 'Angkar' mean for you? Is it an ordinary word or does it evoke fear?"

"The term 'Angkar' was just an ordinary word used at the time. I wasn't frightened to use it, because it was widely used."

"And the word 'pity,' was that used?"

"I never once heard the word 'pity' used at S-21. Not once."

"Did prisoners ask you for help?"

"Yes, they asked me for help. But I told them I couldn't. It wasn't up to me."

"And was that an ordinary job?"

"I only remember some parts of the job. I don't remember the details."

"You spent almost four years at S-21. Are the memories you have ordinary or painful ones?"

"I suffered during my time there, but I had no choice. I couldn't run away. I didn't realize that the regime was exterminating a large part of the population. I was just trying to survive."

"Are the memories painful because you suffered, or because others did? Or was it just ordinary suffering?"

"The suffering at S-21 was immense, because we had to work hard. We had no choice."

"Was the suffering worse for you or for the prisoners?"

"The prisoners suffered more than the staff."

The daily tedium of the trial lulls its participants into forgetting the magnitude of the crime. But after hearing four former S-21 officers on the stand, Judge Lavergne is left seething by the way their testimony reinforces the banality of evil. Four months into Duch's trial, the judge continues to guard against any slump in the collective sense of outrage. We're told a process of dehumanization was required, to enable such crimes. The judge wishes to ensure we remain emotionally invested, if restrained, throughout the trial. The defense lawyer's job is to make us see Duch's humanity and thus underscore his potential for rehabilitation. Regardless, the judge insists that the trial's moral compass remain the solemn and uncompromising refusal to accept the transgressions that took place at S-21.

Each person at the prison had his own, strictly defined tasks. The warden Him Huy's relationship with the interrogators was not a close one. He oversaw the officers who guarded the cells, but not the prisoners themselves. When an interrogator wanted one of the detainees, he gave the prisoner's name to Suor Thi, who would tell Him Huy in which cell and in which building the guards could find him. In return, the interrogators informed Suor Thi in which individual cell the prisoner was to be kept during his interrogation. Once it was over, the interrogator sent the prisoner back to the group cell without going through Suor Thi.

"Whenever a prisoner died in the cell, I received a medical report and then made the necessary adjustment to the list," explains the dull, conscientious bureaucrat Suor Thi.

> When the medical unit wanted blood from prisoners, they put in a request to Hor, who asked Duch, since no prisoner could be removed without Duch's authorization. I did not personally witness any blood-taking, but all those prisoners who had blood taken died. Hor then received a report with a list of names from the medical unit. I checked those names off my list and that was the end of it.

Suor Thi's testimony triggers a gasp in the gallery, but the soundproof glass wall separating the former registrar and the tribunal from the public prevents the court from hearing it.

The media go after the interrogators and the guards, because theirs is the raw story; they can provide the narrative and imagery behind the brutality and killing. The media is less interested in Suor

Thi, even though he personifies the silent bureaucracy that under-pinned the crime. Suor Thi has the accuracy of an accountant and the cold meticulousness of his former boss, Duch. No one spoke to more people condemned to die than Suor Thi.

**IN ONE OF THE PHOTOS** taken inside one of the group cells, you can make out a tangle of men lying on the floor in the background. One of them, his arm folded beneath his head, seems to have a blanket. In the foreground, another man, in a shirt, is sitting up and looking at the lens.

"I don't see how that man could be sitting. It wasn't allowed. Even crying wasn't allowed," says Chum Mey, flushed with anger.

Along with Bou Meng and Vann Nath, Chum Mey, aged seventy-six, is the third and final still-living survivor of S-21. In the court-room, he stands up and, with his hands pressed together in front of his face in *sampeah*, the traditional Khmer greeting, he turns toward the Buddhist monks sitting in the front row, then to the rest of the public gallery, which on this particular day is filled with students. When he tells the story of the exodus of April 1975, during which his two-year-old son died, no detail seems to him too trivial. Chum Mey has repeated his story so many times over the past thirty-five years that he remembers even the smallest detail. The spirited, intense, uninterrupted flow of his speech contrasts sharply with Vann Nath's sober and sparse monologue. When he reconstructs the terrible days he spent in the individual interrogation cell, Chum Mey stares ahead, his eyes filled with pain. When he describes how thin he was, his voice rises at the end of his sentence to a pitch so high it sounds like a so-prano's vocal scale. He describes hearing voices shouting at him: "You sons of bitches, the Angkar will destroy you all! Don't worry about your families!" His hands were tied, his eyes blindfolded, his ankles in chains.

Chum Mey remembers sitting in a room into which he had been dragged by the ear from his cell. Someone took off the blindfold that had covered his eyes from the cell to the interrogation room. He saw fresh blood on the ground. His interrogators asked him how many people in his network had joined the CIA and the KGB. Chum Mey had no idea what the CIA was. Or the KGB. He had heard those terms before, but he didn't know what they meant. For the Khmer Rouge, the enemies of its enemies were also its enemies. Thus, they denounced both the American CIA and the Soviet KGB. The Americans were the imperialists *par excellence*, of course. But the Soviets and their Viet-namese allies were dangerous, expansionist reactionaries with whom the Chinese and their Khmer Rouge allies were engaged in a struggle for the mantle of international Communism. As the "highest tower of proletarian truth," the Communist Party of Kampuchea considered Vietnam and the Soviet Union like "bones stuck in their throat that had to be removed," as Duch wrote in a letter to a very high-ranking prisoner.

All this gave a humble handyman like Chum Mey a great many enemies to learn about in short order. Formerly a tractor mechanic in Phnom Penh, he had been working maintenance in a clothing factory when he was arrested. He tried to show deference to his captors by using reverent forms of address, even calling them the Khmer equiva-lent of "sir." He got a hundred lashes for his efforts. He was to address them as "brothers." Then Comrade Hor, Duch's deputy, rolled up his sleeves and began beating him with a stick. When Chum Mey tried to protect himself, they broke his fingers. Then they clamped elec-tric wires to his earlobes. The wires ran not to a manual generator, as Duch has claimed, but directly from the wall sockets to Chum Mey's ears.

"Kyoukyoukyoukyoukyou . . ." emits Chum Mey, smacking his tongue against his palate in imitation of the electric shocks, before miming how his eyeballs jumped out of their sockets. "Duch didn't beat me personally," he says. "If he had, I doubt I would ever have seen the light of day again."

No prisoner at S-21 was addressed by his name alone. Instead, these survivors were addressed as *a*-Meng, *a*-Nath, or *a*-Mey. In Khmer, to use the diminutive prefix *a-* when addressing a man is a sign of contempt. *A*-Mey describes again how he was called a son of a bitch. The presiding judge asks him to watch his language. *Tell us about the torture but don't swear* is the message.

Chum Mey was beaten and insulted for twelve days and twelve nights. A guard sat on his head. His toenails were torn out, a process that took two days. Presiding Judge Nil Nonn won't tolerate swearing but doesn't flinch at gory details.

"Were the toenails ripped out entirely, or just partially?" he asks. "Did they grow back?" The survivor is now standing in the middle of the courtroom. Someone asks the cameraman to zoom in on his feet, and in an instant, the courtroom is a circus. It doesn't matter which courtroom I'm in—Arusha, Freetown, or Phnom Penh—I've seen it happen time and again.

Finally, Chum Mey confessed. The torture came to an end. Before his imprisonment, he hadn't known that the CIA and the KGB had even existed. Now he admitted to working for them both. What's more, the pain brought on by the beatings had become sufficiently unbearable for him to remember the names of his many accomplices. He named sixteen of them. No one could save anyone else, he says. It was every man for himself: "They just told me to think about my network. I didn't have much time to respond. I don't know if others had denounced me, if that's why I was arrested."

It's more than likely. Chum Mey's boss, the director of the factory where he worked, was arrested before him. His deputy followed. Then another employee. Then Chum Mey. They were all dots on one of those "lines" that had to be erased.

IN A HOUSE SITUATED behind Building A, Bou Meng was forced to lie on his stomach. The windows and door were shut. Someone asked him to pick the stick with which he'd like to be beaten.

"I said, 'That's for you to choose, brother.'"

It was the chief interrogator, Mam Nai, a man with pimply skin covered in red rash, who started, says Bou Meng. Then someone else took over. They heaped insults on him. They made him count the strokes. When his lacerated skin began to bleed, someone threw salt water on his wounds. Sometimes, as many as five interrogators unleashed their fury on him at once. They threw jackfruit skins at his head. After the beatings, they sometimes gave him tablets, so-called medicines produced by the Glorious Revolution. Bou Meng calls the medicine "rabbit pellets."

Bou Meng raises a hand to his forehead, looks up to the ceiling, pulls out a handkerchief, continues. Duch sits perfectly straight in his chair, his attention unflagging, his eyes fixed on the survivor. He drinks more water than usual.

"I didn't know anything about any CIA or KGB networks. I didn't know what to say. I would be happy if 50 percent or even 60 percent of justice had been done. Because I had committed no offense," says a visibly upset Bou Meng.

For thirty years, this thought has been boring through his brain like a gimlet: what crime had he committed? He had done every single thing that the Angkar had asked of him. Yet his torturers told him again and again that it was pointless to ask himself such questions, since the "Angkar was like a pineapple"—it had hundreds of eyes that saw everything, everywhere.

One day, Bou Meng was taken to Building D. He was electrocuted until he lost consciousness. They woke him by throwing water on him. But he still didn't talk. So, weary of waiting, his tormentors ended up writing his confession for him. All he had to do was sign. Remembering this, he says he would've preferred to die of malaria in the jungle. Then he massages his forehead with Tiger Balm:

> I don't remember what was in my confession. I had conflicted feelings at the time. I was fearful, I was worried, I did what I was ordered to do. They had absolutely no reason to suspect me of being a CIA or KGB agent. I signed with my hand. But in my heart of hearts, I did not confess.

# CHAPTER 5

**P**RAK KHAN HAS THE AMIABLE FACE OF A TOUGH AND SELF-assured older man and the thin, delicate lips that so often grace the Khmer people. You can see the muscles of his strong back through his jacket. At fifty-eight, his hair has yet to turn gray. He lost an ear to the war. Like Him Huy and much of the staff at S-21, he was a soldier in the 703 Division of the Khmer Rouge Army. Then he served first as a guard outside the prison—near the canal on Street 360, where Vann Nath was driven before being escorted on foot into the prison proper. Next, possibly toward the end of 1976, Prak Khan became an interrogator. He learned how to interrogate on the job, by watching others and attending the training sessions set up by Duch and run by his deputies, Mam Nai and Pon. Prak Khan learned what "CIA" and "KGB" meant—like most of his victims, he had never heard of them before. In court, he admits that he still wasn't completely clear who the enemy was, even after the training sessions.

He learned the procedures, politics, and the use of torture. The goal, he explains, was to know how to inflict pain on a person without killing them. In theory, at least, the interrogations were strictly regulated. Beating someone to death was forbidden. Some of the techniques taught included electrocution, whipping, asphyxiation with a plastic bag, and inserting needles under the victims' nails. Waterboarding and venomous insects weren't on the list, says Prak Khan. "We told the prisoners not to make any noise, not to swear, and not to cry out while we tortured them."

The objective was to learn the names of the other traitors in a given network. Prak Khan dealt with low-level prisoners, not those that the Party considered important. The interrogators were divided into "cold teams," "hot teams," and somewhere in between, "chewing teams." The "cold" interrogators obtained confessions through questioning alone.

"The cold method was to understand the person," explains Duch. "I don't believe the other interrogators pursued this method to the extent that I did. In principle, according to my training, interrogators had to try to persuade the prisoners first by speaking with them before actually resorting to torture. But often they focused more on the torture."

The "hot" interrogators used electrocution and any other cruel methods they found effective. As for the "chewing" team, says Prak Khan, who was a member, their instructions were to "interrogate in-depth."

But Chum Mey dismisses all these distinctions: "Torture is always hot. It is never cold or lukewarm, it's always hot."


**DUCH FIDDLES WITH A YELLOW PEN.** He seems serene. Though the court is in recess, he remains in the courtroom, making notes and underlining passages. He looks up, flashes a sparkling smile, and starts a conversation with the Cambodian assistant to his defense team. After about ten minutes, he slips out. The filmmaker Rithy Panh enters the public gallery and sits in his usual place on the left side of the front row. The trial is being broadcast on the screen nearby. Rithy Panh watches carefully. The film made by the court is of very poor quality and has become an ever-growing source of irritation to him.

Outside the courtroom, the tribunal's Public Affairs Section tries to satisfy the media's appetite and offer it more scintillating distractions. Sometimes this leads to blunders—for instance, the day they put a cousin of Pol Pot's, who had been bused in to watch the trial, in front of journalists during the lunch recess. The Public Affairs people had promised the media Pol Pot's brother. But the brother wasn't there, so they threw this cousin to the lions. She had never even met Pol Pot, a.k.a. "Brother Number One."

"I believe this court will do justice to those who perished," she said meekly.

Rithy Panh was fuming.

"What was Pol Pot's favorite food?" he said bitterly.

The impromptu press conference quickly degenerated. The relentless pressure from the microphones and cameras pushed this peasant woman to tears. She felt completely lost, she said.

"This is a disgrace!" thundered Rithy Panh. "You've reduced this poor woman to tears, and she didn't even know Pol Pot! Are you mad? Let her go home!"

Duch has now returned to the courtroom, his hands in his pockets. He starts talking to his defense team. A judge walks in without his robes, the garment that elevates lawyers to demigods. He summons one of his legal assistants, then leaves the courtroom only to return a moment later wearing his hallowed garment. The trial resumes. The villagers return to the public gallery. The guard on duty today—his shirt the shade of UN blue, intended to reflect the organization's mercy—wakes two dozing peasant women. They laugh with a reassuring lightness when he gets their attention. The psychologist who examined Duch is sitting among them. She is here to update her assessment. Duch stares at her through the glass wall. His piercing gaze wavers. He snaps out of it and, relaxing his face, brings his attention back to the proceedings.

"Did you see Duch torture a female prisoner?" a judge asks Prak Khan.

"I didn't see it clearly. I don't think he did. He just interrogated her. Others tortured her. Dek Bou beat her and electrocuted her. He suffocated her until she passed out."

Without taking his eyes off the witness, Duch leans forward and pours himself a glass of water. He is fully present now; he has recovered his usual intensity.

"Sometimes Duch would come by to ask if the prisoner had confessed yet," says Prak Khan.

An interrogator would never draft an incomplete confession, explains the former interrogator. A confession was deemed suitable only when the interrogator had clearly identified a network of traitors. Once he had his list of people, he gave his report to his superior, who checked it, then passed it on to Duch, who, like the schoolteacher that he is, made notes in the margins in red ink. Once the confession was deemed complete, the prisoner was taken from the brick-walled isolation cell back to one of the group cells, where he would remain until his execution. According to Prak Khan, more than half the prisoners were never interrogated. They went straight to the killing fields.

When Prak Khan says that he could read a little French at the time but that he has forgotten it since, a smile creeps across Duch's face. He smiles often during his former subordinate's time on the stand, his expression turning into a muted leer of condescension whenever Prak Khan displays his limited education. And when the witness's story veers into implausibility, Duch looks up at the ceiling and smirks in disbelief. Whenever he lets down his guard, Duch finds it difficult to control his laughter. As the next recess is called, Duch stares at Prak Khan awhile, obviously resisting the urge to laugh, before getting up and leaving the courtroom.

Along with Him Huy, Prak Khan is one of the greatest threats to Duch's case. Duch's response is to eyeball them both in the courtroom, to carry out a silent campaign as fierce in its intent as their testimony against him is devastating. Waiting for the judges to return at the end of recess, Duch stares at the former interrogator, a wry smile on his face. Prak Khan doesn't dare meet the gaze he surely feels against the back of his neck like a cold blade. Duch lets his eyes wander away, and looks over the vast and packed public gallery.

Prak Khan and Him Huy have been telling their stories for years. To this day, they agree to virtually every interview request, which means that for three decades, their memories have been spurred and goaded to the point that they almost certainly generate many speculative and spurious claims alongside the verifiable ones. One of the law's most common—and most mutable—tasks is to separate facts from fictions that inhabit our memories. The conflicted and incriminating memories of torturers and executioners are trickier still.

A case in point: prior to Duch's trial, Prak Khan said that he had seen Duch personally administer electric shocks to prisoners, which is the only such evidence against the defendant. But Prak Khan won't repeat it in court. Instead, he either shifts the blame onto another interrogator or else claims he can't remember. And indeed, by the trial's close, the resolution of the most serious allegations leveled directly at Duch—including Him Huy's testimony about what took place in the killing fields of Choeung Ek and Prak Khan's claims of what went on in the interrogation rooms of S-21—appears to depend more on personal conviction (I believe / I don't believe) than on the legal burden of proof (it happened / it didn't happen).

Him Huy and Prak Khan risk nothing: they won't face trial. But, like Duch, they're trying to untangle their memories.

"I didn't participate in torture, but I saw other prisoners tortured until they passed out," says Prak Khan.

It's always somebody else; somebody who's now dead.

The trouble is that Prak Khan has previously admitted that "chewing" included torture. He has also admitted, though not in court, that he himself tortured prisoners by electrocution, by beatings, and with animal traps. A judge reminds him of this: "Do you wish to comment, or do you choose the right to remain silent?"

"I do not want to add anything. It reflects the truth."


**THE TRUTH IS A HARD THING** to articulate, let alone take responsibility for. Though we have good reason to decry the torturers' failure to speak the truth, the temptation to omit and obfuscate the unpalatable preys on us all. Who among us has not been tricked by his own memory or perception? In court, not even the victims speak "nothing but the truth." No one ever speaks "nothing but the truth."

A courtroom is a place of high drama and strong emotion, where truths emerge and stories reach their dramatic climaxes. But it can also be a terribly sterile place and the source of great disappointment. With Prak Khan, it's the latter. In Rithy Panh's famous 2003 film, *S-21: The Khmer Rouge Killing Machine*, we see Prak Khan talking about drawing blood from prisoners; we hear him describe how he considered the detainees to be nothing more than animals; we hear how he disassociated himself from his actions. Compared to what he said to the camera, the testimony he gives in court is dismayingly benign. The terrible allegations he made on film are stifled in court, hidden behind his right to remain silent. In a canny move, Duch's defense team has threatened legal action against former S-21 personnel. That they could pursue such a course is highly unlikely, but the threat alone is deterrent enough. Which truth, then, should we believe—the legal truth or the filmed one? Some of Prak Khan's and Him Huy's "memories," either imagined or based on hearsay, vanish the moment they take an oath to speak the truth.

Duch denies Prak Khan's assertion that he regularly attended interrogations. He reminds us that everyone adhered to the strict hierarchy he had instituted at S-21. The clerk Suor Thi was very much at the prison's nerve center, for example, yet he had no personal contact with Duch, its head. To get a message to Duch, he had to pass through Hor, the second-in-command at S-21. Suor Thi never received a direct order from Duch—whom he called "Brother East" because Duch was situated on the east side of the prison. Suor Thi feared Hor, "Brother West," but he also knew that Brother West was afraid of Brother East. What he didn't know was who Brother East, a.k.a. Duch, was frightened of.

"That was beyond what I could know," he says.

Like the prison's other officers, Suor Thi regularly saw his colleagues get arrested and disappear. Fear was omnipresent. Everyone was terrified. More than 150 S-21 officers were victims of purges at their own prison.

Prak Khan's lower rank prevented him from reporting directly to Duch, explains Duch. He had to go through an intermediary.

Duch says evenly:

> *He couldn't go over his supervisor's head to talk to me. If I had discussed the slightest thing with Prak Khan without going through my deputies, what was the point of having them? I didn't have the time to give instructions to each individual interrogator, including Prak Khan. I had never met [him] or even heard his name until January 7, 1979. He was a low-level member of staff. I spent some time trying to understand his life story.*

Duch stands when addressing the court. His face is gaunt and he looks old and tired. He summons his energy, then says in a conciliatory tone:

> *I believe Prak Khan's testimony, wherever it leads. There's a lot in it that's false, but I think this is the result of fear. Back then, you were afraid that I would have you arrested. Now, like me, you are afraid of having to face the tribunal. But I neither wish for nor need my subordinates to appear by my side before this court. I am responsible for what happened at S-21.*

At this moment, an old revolutionary habit catches Duch off guard: he raises his hand in a military salute. He goes on to list the errors and extrapolations in Prak Khan's testimony, before admonishing: "Never say anything without material proof! You are making subjective claims without any supporting documentation."

Presiding judge Nil Nonn calls Duch to order. Judge Cartwright pulls out a document. It's dated February 1976 and comprises the minutes of a meeting of Phnom Penh's defense force, when Duch's job was to teach Party doctrine. The document quotes him as saying, "Forget the idea that beating a prisoner is cruel. There's no place for kindness in such cases. You must beat them for national, international, and class reasons."

Duch counters with another set of minutes, written five days earlier and in which this damning statement doesn't appear. Even with supporting documents, the torturers' truth is slippery to the point of being exasperating. When the court asks Suor Thi to explain a document in the archives for which he was responsible, he says that the page isn't formatted the way it was back then, or the number of columns is wrong, or it was written on a different typewriter, or the annotations aren't correct.

**HIM HUY, A STILL-SPRIGHTLY OLD MAN** with laughing eyes and thick black hair, appears in court wearing an apple-green shirt over a yellow T-shirt, his spectacles tucked into his breast pocket. The former guard has a face shaped like a mango and long, beautiful, pointed lips. He suffers from a nervous tic that makes him sniff and blink continually. Him Huy is typical of the indoctrinated youth who worked at S-21— the young peasants recruited as henchmen by the teachers who ran the complex. He was an illiterate seventeen-year-old when the Party enlisted him. He soon found himself under the command of one of Duch's two deputies, the number three of this little archipelago of death, who ran the "reeducation" center S-24.

Conscripting children into war or revolution is nothing new. Duch recruited teenagers because they were

> like blank pages on which you can easily write or paint. . . . We took
> in many young people and trained them to be cruel. We used Communist
> jargon to normalize extreme situations—that played a big part in
> turning innocent people into brutes. Their characters changed. Their
> kindness gave way to cruelty. They became motivated by class rage.

When Him Huy relates how his fellow officers were arrested, Duch smiles, puckers his lips, and raises an eyebrow, as if to say that the story strikes him as not quite true. Duch is so relaxed he comes across as almost arrogant. He exudes a quiet pressure in the courtroom, and makes his presence felt. Him Huy sniffs. He describes how the purges at S-21 threatened the most senior officers, including Duch's two deputies. He explains that, from 1976 on, he was promoted after each purge until eventually he was a head guard. When Nath, Duch's predecessor as the head of S-21, was crushed by his own machine toward the end of 1978, followed shortly thereafter by one of Duch's deputies, Him Huy worried that he would be next, that he was about to be swept away along with the rest of one of those "lines" that the Angkar traced between people before wiping them out. He sniffs again: "Honestly, when I see him now, it reminds me of when I worked for him and was frightened of him. He still frightens me. If we hadn't been liberated on January 7, 1979, I don't think I would be around today."

Him Huy slumps into his seat and pulls out a big yellow handkerchief. "I would've been killed because Duch said so. He said everyone would be killed in the end."

Suor Thi describes his terror at the end of 1978, when the leaders started wiping one another out. All his hard work had amounted to nothing, he tells the court. "All I got in return was fear."

"Did you like your work?"

"Not for a second. I hated it, but I had to do it."

"Why didn't you leave?"

"Where could I have gone, under the regime? If I was five minutes late, someone noticed. I had no choice but to force myself to do my job. There was nowhere to escape to. The personnel at S-21 didn't like the regime, and that's the truth."

Before stepping down from the stand, the S-21 survivor Chum Mey has some questions he wants to ask Duch. When he was accused of being in the CIA, were all the agency's agents eliminated or did a few remain? What exactly was the Angkar? And was Pol Pot the same thing as the Khmer Rouge? All these questions are still on his mind, he says. He would like to have clear answers for the schoolchildren to whom he speaks from time to time.

Head Judge Nil Nonn looks amused by the old survivor's naive questions. Duch smiles. There's nothing he likes more than wearing his teacher's hat.

"I want to make it clear that the term 'CIA' referred to those people who opposed the Communist Party of Kampuchea," says Duch. "The real CIA and the Party's CIA were two different things."

As for the Angkar, he says, it was basically the Party's permanent committee. But the concept, by its very essence, had to remain mysterious.

For Chum Mey, the memory of all those he denounced in his confession is a painful one. He loses his temper. Again and again, he goes over the accusation of treason that had been leveled against him, an accusation he had never understood, that had never made any sense to him and yet had cost him so much. Behind all of his questions to the executioner lies the enormous, unyielding stupefaction of a sane man vainly seeking to understand the inexplicable.

# CHAPTER 6

**D**UCH STANDS AND GREETS THE COURT. HE BEGINS BY SAYING that his people's suffering started with Prince Sihanouk's repressive government in the mid-1960s and continued after the far right–wing coup of March 18, 1970, when "all the parties competed to kill Cambodians until April 1975." He holds a sheet of paper in one hand and leans on the edge of the table with the other. It takes just seconds for the room to fall silent. Though the trial started the previous day, not until now has it been imbued with that solemn atmosphere so specific to important moments in courtrooms. Duch is asking for forgiveness:

> *No single image can illustrate my remorse and suffering. I feel so much pain. I will never forget. I always say that a bad decision can lead in the blink of an eye to a lifetime of grief and remorse. I defer to the judgment of this tribunal for the crimes that I have committed. I will not blame my superiors. I will not blame my subordinates. I will not shirk my responsibilities. Although these crimes were committed under the authority of my superiors, they fall within the purview of my own role at S-21. On the ideological and psychological levels, I am responsible. I carried out Party policy and I regret it.*

Bou Meng nods approvingly. Duch looks like he's trembling. The judges barely look in his direction. He removes his spectacles and leans on the desk with both arms. He looks at each person in turn, first left, then right, giving most of his attention to the prosecutor's bench.

"I never liked my job," he says.

When he describes his arrest in May 1999, Duch's breathing grows heavy and he sounds ill. He finally mentions the sheet of paper he's been clasping since the start of his address to the court. It's a drawing that he has made, he says. He would like to show it to the judges. He sits back down while waiting for their permission. People have been waiting thirty years to hear Duch speak out in his own defense. The public gallery is abuzz. Yet the moment is utterly devoid of emotion.

Duch tries to explain his peculiar drawing. He points to three chairs on the sheet of paper, which he says are occupied by Pol Pot, Nuon Chea, and Ta Mok—Brothers Number One, Two, and Four of the Khmer Rouge leadership. Along with Brother Three, Ieng Sary, this was the structure of the Communist Party of Kampuchea, explains Duch. The presiding judge keeps his eyes glued to the defendant. The other judges look away. Duch's first address to the court is a resounding flop.

AFTER ARRIVING IN PHNOM PENH on June 21, 1975, Duch, like everybody else, went through a few days of political training. He was taught "the revolutionary conception of the world," he says in French. Each person was made to write down his "biography" and ideology. Celebrating the "great victory of April 17," the date the Communist insurrection took Phnom Penh, was mandatory. So was committing oneself in writing to the good of the collective, to the teachings of socialism, and to the continuation of the Revolution. Once a person had written down his biography and commitments, he read them out to his comrades, who were then encouraged to ask questions. He also had to reveal his family background, which was far more perilous than it sounds: having the wrong family tree could get you killed. Duch made sure not to mention that he was related to the niece of Lon Nol, the recently deposed field marshal with a price on his head.

Duch says that it was at this time that he tried to quit the Party's security services after having worked for them for four years. He asked a high-placed contact for a transfer to the Ministry of Industry, he says. When the court asks him to elaborate, Duch answers with a proverb that the judges, particularly the foreign ones, are free to interpret however they choose: "Is it necessary to crack open a crab to see its shit?"

When asked whether he hadn't developed a taste for police work, if he hadn't found fascinating the secret and all-powerful world of the Party security apparatus, Duch has no good answer. Pressed, he dodges the question. Pressed further, he rehashes the explanations that his conscience has already endorsed: that his work was evil by its nature, or that the confessions were half-false. But the difficult question of whether he enjoyed committing the crimes won't go away, and eventually Duch makes an effort to answer it.

His effort fails.

"I was just an instrument of the Party," he says, defeated, "an absolute, authoritarian instrument."


**THE S IN S-21** stands for *Santebal*.

In the Buddhist lexicon, the Santebal are those who keep the peace and maintain order, like the police. Under Pol Pot, Santebal was the name given to the internal security service, more commonly known in Communist regimes as the secret police. At the end of June 1975, Son Sen, the head of state security and minister of defense, informed Duch that a detention and intelligence center was being created in Phnom Penh. He told Duch that they were to follow the French "counterespionage" model.

The *21* in S-21 was, according to Duch, the radio code that belonged to the center's first director, Nath.

Duch was sent to search government buildings and the homes of former government employees. He gathered reports and archives from the fallen regime. From the judicial police headquarters, he took documents on torture. S-21 was created on August 15, 1975, with Nath as its director and Duch as his deputy.

It was set up first and foremost to eliminate the *ancien régime*. This included army officers, civil servants, aristocrats, and "new people"—those who stuck with the old regime right to the end and those who lived in the cities. The revolution soon found that it had no use for the mentally impaired, either: in its earliest days, S-21 served as a psychiatric hospital as well. What became of its patients?

"Based on my own analysis, more than 50 percent of patients were smashed,* though I'm not entirely sure," says Duch.

He has a better recollection of what he was ordered to do with lepers: destroy them all. Communism must liberate man. Communism abhors the handicapped, the sick, the mentally ill, the religious, homosexuals, and intellectuals.

On March 30, 1976, Party leaders signed a secret order authorizing purges within the Party itself. It would prove a watershed moment. That decree is the most tangible proof we have of the policy of extermination implemented by the secretive Angkar. The order formalized as policy the already existing practice of summary execution by giving the zone committee, the central committee, the standing committee, and the military staff the authority to kill. Thus began the great purges, ministry by ministry, division by division, region by region. Nath lost his job, Duch was promoted, and S-21 had a new mission. Its focus was now on the internal purges, as per the decree of March 30. Yet Duch was unaware that this decree even existed. He would only learn of the Angkar's decision some thirty years later, while in prison.

"Why were you chosen to run S-21?" asks Judge Lavergne.

*First, it's true that I was a much better interrogator than Nath. But it was more than that. The Party had no confidence in him. Son*

---

\* "To smash" was the official English translation used in court for executing or killing people. For the sake of accuracy, I have kept it throughout, whenever it is used in a quote.

*Sen used to say that Nath's methods were dubious and that he was a schemer. I was honest. I would have rather died than lied to a Party member. And I was loyal. I reported every single thing I learned. I was methodical about it. All my life, whenever I've done something, I've done it thoroughly.*

Duch claims to have been terrified when he took over. He says he even suggested that someone else take the position instead. But Son Sen threatened him, he says. When he tries to reenact their conversation in court, the pitch of his voice climbs until his Khmer sounds metallic, jarring.

"I was their sheepdog," he says.

But for the prosecutor, he was clearly the perfect fit for the job.

**THE HISTORIAN DAVID CHANDLER** likes to say that S-21 was probably the most efficient institution in all of Cambodia during the Khmer Maoists' tragic and grotesque reign. Its own impeccable archives showed that S-21 was efficient, modern, and professional. The archives, down to the smallest detail, convinced the Party leadership that its suspicions were well-founded. They satisfied the Party's need to prove that it had eliminated all its enemies and that it had emerged victorious, even if newly conjured enemies constantly surfaced.

Chandler doesn't think the Khmer Rouge followed any particular Communist model. Similar security centers existed in China and the USSR, where the security apparatuses extorted the most dubious confessions and "reeducated" reactionary minds with the same unabated enthusiasm as the Khmer Rouge did at S-21. From Lenin onward, the Russian Revolution was blighted by purges. What Chandler does believe is unique to S-21, however, is its completely secret nature. And he calls the practice of "reeducating" prisoners only to then kill them unprecedented. The systematic killing that took place at S-21 made it a unique combination of a secret police prison and death camp.

A network of prisons and interrogation centers in which black-clad agents carried out violent abuse abounded across the Khmer Rouge's Democratic Kampuchea. But each little island in this police archipelago was isolated from the others. S-21 had no authority over any other prison, nor any autonomous or direct contact with them. Everything had to go through the "center." S-21 was unusual because it did have a sort of national jurisdiction in that it could receive prisoners from anywhere in the country, and because it was directly linked to what everyone called the "upper echelon." S-21 was an arm of Santebal directly linked to the center of power: the standing committee, the true Angkar, which comprised between five and seven members. This prison was its exclusive tool. The most important arrestees were sent here and nowhere else, including those made within the Central Committee or the Politburo.

But S-21 is also unique in that we have its archives. It is often emphasized that no other prison in Democratic Kampuchea was run as efficiently or with such sophistication. Perhaps. But no other prison in Cambodia remains with its archives intact. We know little about the two hundred other centers of the secret police that have been identified, just as we would know nothing about S-21 had Duch been ordered or had the presence of mind to destroy its records.

Duch's confession makes him unusual among the members of the Khmer Rouge's inner circle. But would he have confessed had he not left so much evidence behind? Duch is a mathematician; his arguments adhere to logic. He has admitted nothing that can't be found in the archives.

"If documents exist then I can't deny it," he says simply. "I recognize everything that comes from S-21. I accept no other evidence."

His first great error was not anticipating the rapid Vietnamese invasion in early 1979, and his second was not destroying his archives before fleeing. His superiors made the mistake of not ordering him to do so. Didn't they know about their warden's meticulous record-keeping? Did Duch leave the fruits of his endeavors intact out of haste or simply out of lack of foresight? Or was it because he had been too proud to destroy the exemplary work he had accomplished for the

Party and for the Revolution, the testament to his talent and proof of his ability to establish a successful and efficient institution?

Had Duch destroyed his archives, we never would have known much about the prison nor the magnitude of the crimes committed there. We might never have known the true identity of its director, "Brother East." It's true that a handful of survivors might have told us about a terrible place that had once existed in Pol Pot's Cambodia. But with no written records, how many other terrible prisons have been erased from the pages of history? Without the written confessions, photographs, and "biographies," there is no S-21. In short, S-21 exists today because hubris or professional oversight prevented its director from destroying his work.

The archives are of exceptional quality and incredibly thorough. Without them, the history of Democratic Kampuchea would be much murkier and less detailed. And if we had all the regime's documents, says Chandler, "We would have a completely new history of Democratic Kampuchea." Ever the iconoclast, Chandler tells the court:

> Maybe S-21 was not as important [to the Angkar] as it is to those of us seeking evidence about the Democratic Kampuchea regime. I think if we had [all the minutes of Angkar] cabinet meetings, I'd be very surprised if S-21 gets mentioned very often. Certainly the top leaders were very interested in the confessions of high-ranking cadres, but for the people who were not high in the chain of command, [they] would not be interested.

# CHAPTER 7

**D**UCH WAS NOT A MEMBER OF THE PARTY'S CENTRAL COMMIT-tee, which means he was among the middle rungs of its hierarchy. He was, however, high up in the secret police—a commissar, he says, returning to orthodox Party jargon, at the head of its most strategic and sensitive department. With feigned modesty, he describes his position thus: "I had three important jobs: to teach and train; to send confessions to my superiors; and to deal with any questions having to do with S-21. My duties were more political than technical."

According to Chandler, Duch was an outstanding administrator who expertly accomplished the work asked of him. Little of the prison's operation escaped his attention. Yet even so, says the historian, not even he could instigate and manage everything that happened there.

Duch maintains that he went infrequently to the prison, and never into the cells. He left the menial tasks to his subordinates, he says, taking care to cultivate the mysterious and menacing persona of the inaccessible boss who appears rarely and then only by surprise. He devoted himself entirely to his strategic role: reading and annotating confessions, sending them back to the interrogators with comments, then reporting on them to his superiors so that they would "understand them more easily." He did not wish to see the conditions in which the prisoners were kept, and in any case he paid them no attention whatsoever. This is what psychologists call "avoidance."

"S-21 was only for people who were going to be executed. There was no protection of their rights. We fed them like animals and treated them as such. We were only waiting for the moment when they would be smashed. No one cared about their well-being. That's it."

"You admit that you didn't consider them human?" asks a lawyer for the victims' families.

"We didn't think about it in such complex terms. We distinguished between friends and enemies. Now, looking back through the prism of human rights, it was absolutely wrong and constituted a criminal act. But at the time, we thought about it differently. We told ourselves only that the police work had to be done," says Duch.

Duch oversaw the logistics of death, but from a distance. He left the daily management to his deputy, Hor, who had an office inside the prison, where the clerk Suor Thi worked.

Apart from closely monitoring the confessions, Duch's other tasks were to train his staff in interrogation techniques and to ensure their political education. The crimes perpetrated at S-21 were the work of many, and though Duch certainly got his hands dirty, his real aspiration was to become a bureaucrat. To a certain degree, he succeeded.

**OF COURSE, HE KNEW EXACTLY** what was going on. He is an expert on torture. The method of covering someone's head with a plastic bag is "very dangerous," he asserts. Throwing water over it makes it even more sophisticated. The prison's first director, Nath, had a predilection for electrocution and whipping. Duch says he himself authorized four types of abuse: forcing water up people's noses, beating people up, whipping, and electric shocks. He doesn't believe that prisoners had their breasts burned, or that his interrogators used venomous insects such as centipedes to extract confessions. Above all, he tries to convince the court that he did not pay close attention to the technical aspects of what went on at S-21; he left that to his underlings: "I didn't know what they were doing and they didn't know what I was doing."

It's a stance that makes it easy to avoid remembering too much. In reality, all kinds of tortures were practiced at S-21. Chum Mey describes having his nails pulled out, though Duch claims he put an end to that method. One written order directs an interrogator to force one of Duch's former teachers to eat spoonfuls of excrement. Brother Number Five, Duch's former boss, was forced to take a cold shower and then sit in front of a fan—"to induce fever," says Duch. Some were made to drink urine. Others were forced to pay homage to an image of a dog with the enemy's face superimposed over it, or made to kneel in front of a chair, or a table, or any other object that only a contemptible person would honor. How should such methods be labeled? Are they cold, lukewarm, or hot?

Duch vows that he never tortured anyone himself. He admits slapping a prisoner around once when Nath was still director of S-21, but he only acknowledges that because he has to—it is documented in the archives. His clumsy explanations are difficult to believe. Even the way he phrases his answers suggests that he knows more than his conscience will admit:

> Usually, torture took place when I was angry. Chet Eav was the police inspector under the old regime, who interrogated Khmer Rouge prisoners; he was very aggressive. Nath wanted to beat him up. He asked me to interrogate him, and Chet Eav finally confessed. I slapped him around to keep Nath from beating him.

"What characterizes torture for you?" asks Judge Lavergne.

"That's a difficult question. Could you phrase it in a different way?"

"Can you imagine how the prisoners felt? The water technique, for example. How do you think that feels?"

"Once the stomach is filled with water, the prisoner is shaken until he vomits and sometimes loses consciousness. When he comes to, we carry on with the interrogation."

He remembers one victim in particular, to the point that he even remembers his name, as well as many details of his interrogation:

> The water wouldn't go through his nostrils. When he arrived at S-21, all sorts of interrogation methods were used on him, but he still

*wouldn't confess. I discussed using other methods with my deputy Hor.*
*I ordered them to try it, but they found out the water wouldn't go*
*through his nostrils.*

"Can you imagine what it feels like to have a plastic bag over your head?"

"It feels like you're dying."

The only rule to the torture—though at times broken—was that the subject had to be kept alive until he or she had made a complete confession.

"Is it easy to know when to stop?"

> *Quite frankly, before I hit anybody I used words. I only started*
> *hitting if words failed. I knew how to control my emotions and my*
> *actions. I knew when to stop. But the young interrogators didn't know.*
> *They were extreme. They didn't have any self-control. Some of them*
> *were crueler than others. The more I think about it, the more I am*
> *moved.*

**NEVERTHELESS, DUCH HAS NO TROUBLE** admitting that he ordered torture. The more specific the documents are, the less he equivocates. When Brother Number Ten wrote in his confession that he had been severely tortured by Duch's men, Duch crossed out the passage in question and wrote to the prisoner: "You have no right to report this problem to Angkar. I am the only one who decides." In another report, an interrogator named Pon made a note of the number of lashes a prisoner had received; Duch wrote instructions to give him more. Elsewhere, he wrote to Pon to torture a prisoner by the hot method, "even if it leads to death." When asked in court what he has to say about this particular note, Duch claims that it was a bluff to frighten the prisoner into confessing; it wasn't for real.

Duch wrote his annotations in a clear, neat hand. They are as elegantly written as they are ruthless.

"I'm very jealous of the neatness of his calligraphy," David Chandler, who has analyzed a thousand confessions extorted at S-21, comments with sarcasm.

Duch wrote instructions to his torturers in a red pen:

> *Did not confess. Torture him!*
> *Hit him in the face*
> *We must apply pressure, absolutely*
> *Beat them all to death*
> *Smash them to pieces*

The radical, uncompromising revolutionary never wavered from inflicting suffering, from destroying men. The young, thirty-three-year-old Duch who reigned over his secret prison and who made these annotations in red ink some thirty years ago, seems to have little in common with the elderly Kaing Guek Eav who appears in court today. He may be a manipulative old man, and he's certainly hiding things, but he's not a threat.

Duch wants to convince the court that, under his authority, torture was at least practiced with a certain amount of objectivity: "I wouldn't say that torture was common. It was carried out only when necessary."

It's true that one of the principal interrogators that worked for Duch, Mam Nai, wrote in his copious notes that putting greater emphasis on torture than on politics and propaganda was an "erroneous position." Mam Nai, a former teacher and a resolute and conscientious subordinate, took notes during the training sessions organized by Duch. To carry out a good interrogation, he wrote, the first step is political pressure, whereas "torture is a complementary measure."

The fact remains that we don't know how systematic the torture was. All we know for sure is that it happened abundantly.

"It's impossible to know how much torture was used and it's quite possible that some confessions reached a conclusion deemed satisfactory with minimal or almost no torture," concludes Chandler.

In 1976, S-21 was moved to the Ponhea Yat High School and Tuol Sleng School, says Chandler, to conceal its existence from the many Chinese advisors then in Phnom Penh. (It is now the site of the Genocide Museum.)

Secrecy: nothing was better maintained than secrecy. So precious was secrecy to the Khmer Rouge leadership that for a long time the Party's very existence was kept hidden, and the name of Brother Number One, Pol Pot, wasn't divulged until more than a year after the 1975 victory, and then only discreetly.

"I was instructed to share nothing with my colleagues," remembers Prak Khan. "I was told to keep everything secret. Each of us had to keep things secret. We were supposed to look after only those things that concerned us, or else we would be reported."

Duch taught his staff that secrecy was the very soul of their mission and that, without it, their work made no sense. Guards and interrogators were not authorized to communicate with other units. Merely having contact with the outside world was deemed suspicious. Secrecy was an obsession, the Party's alpha and omega. It was also a formidable instrument of control that, like everything else in Democratic Kampuchea, eventually imposed its own insane logic over all other lines of reasoning. The systematic execution of prisoners at S-21 was in large part due to the absolute imperative of keeping the prison secret. Due to secrecy becoming of utmost importance, it was decided that nobody could get out alive. And if someone were arrested by mistake, then the secrecy of the institution took precedence over that man's life.

Then there was the fear. Nothing was more widespread than fear. In court, the prosecutor doesn't like it when the defense team emphasizes the atmosphere of terror that reigned over the country at the time. He worries that it's too easy an excuse for the defendant, who, he says, freely chose the path that led to his crimes and who enthusiastically organized the execution of his people.

Nevertheless, under the regime, the threat of annihilation hovered over everyone, as those who worked at S-21 knew better than anyone. Most of the staff working at the prison complex was recruited from a single division of the Army of Democratic Kampuchea, the 703. Both of Duch's deputies, Hor and Nun Huy, were from Division 703, as was his predecessor, Nath. The day the division—like so many others before it—fell from grace, Duch did some housecleaning. Nath's life ended in the prison he used to run. His wife followed him in death. Nun Huy, the third-in-command at the prison after Duch and Hor, was wiped out in December 1978, along with his wife and children, a month before the Vietnamese troops entered the city. Hor, Duch's number two, was also on the hot seat, guilty of having compromised the interrogation of a high-ranking Party official: "The secretary of Division 703 was eliminated first. Then his subordinates were placed under the upper echelon's supervision. They couldn't avoid being purged. It was only a matter of time. That was the process."

Duch had no illusions about his own fate. "I knew it was only a matter of time before I would be arrested."

During the trial's preliminary investigative phase, Duch makes a point of emphasizing the appalling absurdity of the system in which he was caught:

> After each arrest, I would ask myself, "Are they really guilty? Are they really traitors? Am I going to be arrested, too, before I get the chance to know whether it's fair or not?" And I thought, well, I'll just have to wait for them to arrest me before I can dare say that the arrests are unjust.

Yet despite waiting for his own end, Duch's zeal never flagged; he continued to write damning reports about his own subordinates. In no sense could any of them have been described as his protégés.

"Were you aware that your reports meant that these members of your staff would be 'smashed'?"

"I knew that the decision would be made to arrest them," replies Duch evasively.

"That's not a straight answer, but it doesn't really matter," says Judge Lavergne.

Secrecy, fear, and obedience were omnipresent; there was no questioning authority: "I am alive because of my loyal obedience. I never hid anything. Honesty and a commitment to doing things correctly are why I survived. Other survivors probably share these same qualities."

The final quality required by the Khmer Rouge was enthusiasm. There was no better precaution than fervor—a revolutionary cannot, by definition, be lukewarm—and Duch's passion was exemplary. He bent over backward to please his superiors and completed every task assigned to him with the utmost devotion. Judge Cartwright asks him: "People say that you were more enthusiastic than you needed to be. In other words, you did more than was necessary to stay alive. Do you have anything to say about that?"

> What is the norm for measuring enthusiasm? The Communist Party was a paranoid organization. They suspected everything, and they believed that anyone could turn out to be a traitor. There was no criterion for measuring what was within the parameters of an acceptable result and what wasn't. This is my most frank answer, Your Honor.

In psychology, "ambivalence" is a state where two conflicting notions exist within the same person. Not at the same moment (that's ambiguity), but rather almost consecutively. Everyone is capable of it. As one's sense of disgust grows, then, so, too, can one's zeal for the job.

"We cannot live long in an ambivalent situation," the psychologist expert witness tells the court.

> We look for ways out, for ways to adapt. We lean one way or the other. Then denial sets in. It's not conscious, of course. But the zealotry is part of it. That is to say that we can definitively silence that which we cannot accept in ourselves. We search for reasons to silence the shame or disgust. If we can't get rid of the ambivalence, then we risk getting

*physically sick and suffering major somatization, depression, and*
*mental illness.*

Yet that wasn't the case with Duch. He managed to adapt.

**FOR THE PROSECUTION,** the extent of the extermination can't be blamed entirely on the Politburo's policies. Duch made his own contribution.

"One could say that the purges were driven by the methodology used to search for enemies, which was developed and enforced by the accused," says the expert witness Craig Etcheson.

"How was the methodology used by Duch different from the line imposed by the Party? How was it his own personal initiative?" asks François Roux, Duch's French lawyer, who systematically refutes this thread of the prosecution's argument.

> *My understanding is that the accused was very much an*
> *innovator, a creator, a developer, and an institutionalizer of the*
> *method of making very detailed confessions, which were extracted*
> *over long periods. Those lists [of names were] used to go out and*
> *round up new batches of traitors. You see a very nearly exponential*
> *growth in the number of accused traitors and in the number of*
> *victims of purges. In part, it is the zeal with which the accused person*
> *pursued this project that caused this methodology to result in such a*
> *large number of victims.*

"Did he have a choice?"

"One always has choices in life," says the academic, a little haughtily.

"And yet you agree with me that today Duch is still alive?" says the lawyer, clearly very irritated.

"Yes, he is."

The historian David Chandler is more careful. Once the regime of death and terror became a daily reality for everyone living under the

Khmer Rouge, perhaps they no longer had a choice, he suggests. "It's most unlikely that if these decisions were made at the top, that dissent would have come from the middle ranks; it began to roll on once the decisions had been made."

He hesitates, takes a breath, rubs his face.

> *In a way, I'm reluctant to say this, because I've never been in any kind of a situation where I would have been in danger by refusing to do something, but I can't help but think that the people who inflicted this terrible damage on everybody knew what they were doing and, almost worse, did not seem to suffer themselves from what was happening. It didn't seem to cause them to lose sleep. It didn't seem to make their handwriting more unsteady. It didn't seem to lessen their enthusiasm for coming back to work the next day.*

When Etcheson is on the stand, the prosecutor tries to come to the rescue: "I would not wish an impression to be left that there [were] only two choices for a Communist Party cadre—death or duty. Perhaps to complete the picture, the expert should be asked whether he knows of a third way—namely, escape?"

However, under the Khmer Rouge tyranny, only a few thousand Cambodians fled to Vietnam, estimates Chandler. The Vietnamese border was more porous and the refugees were well-received. A few hundred others fled toward Thailand, where they were hardly welcome. All in all, then, only a tiny number of people escaped. And one had to live near a border in the first place.

None of the witnesses state that escape was possible. The country and all of its citizens were being closely monitored. The guard Him Huy, like Suor Thi before him, articulates the twin threats that paralyzed everybody at the time: "Even if I had tried to escape from S-21, I would've been arrested. I was sure of it. Where would I go? To enemy territory? I would've been arrested on the spot. And if I was arrested, or if I fled, what would have happened to my family and loved ones?"

# CHAPTER 8

EVEN BY THE STANDARDS OF A REGIME THAT BANNED FALLOW fields, rest, and recreation, Duch was an exceptionally hard worker. By his own estimate, he sifted through some 200,000 pages of prisoner confessions. He had a peerless command of the many interrogation techniques required of him, including the use of subterfuge, charm, deception, bluffing, intimidation, chastising, punishment, analysis, and summary. Duch saw interrogation as a fascinating and exciting game, even if it was one-sided and fixed. He saw it as an opportunity to apply his intellect, exercise his influence, and win favor with his superiors.

Duch offers the court a glimpse of his skills when he describes the interrogation of Koy Thuon, a very high-ranking member of the Central Committee. First and foremost, he says, it was vital to prevent the prisoner from committing suicide, to which end two guards were posted permanently in his cell. A telephone line was installed so that Duch could be contacted at the first sign of a problem. Duch made sure the prisoner slept well and that he was served the same food as Duch. Upon hearing that Koy Thuon hadn't responded to questions for an hour or two, Duch went to the prison himself and interrogated the cadre leader one-on-one. Rather than use the diminutive prefix a-, which he and the guards used to deride the low-ranking prisoners, Duch addressed Koy Thuon as "Brother." When the mighty fall from grace, they get a little respect prior to being tortured and killed. "Koy Thuon was quick to react. Sometimes he would break a pen or a glass. I'd let him calm down, smile, and say,

'You have no choice but to send your confession to the Party through me.' And he understood."

Koy Thuon also grasped what he would endure if he didn't talk. After two interrogation sessions, Duch handed the prisoner over to his right-hand man, Pon, because he "didn't want to be involved any further." Pon had more persuasive methods, as Koy Thuon would soon learn. Pon was a chief interrogator with an "excellent command" of violence, says Duch. In other words, he was very good at torturing people without killing them. The confession of the great fallen revolutionary Koy Thuon ran more than seven hundred pages.

Once a confession had been extracted from a prisoner, Duch's task was to make life easier for his superiors. To that end, like any good chief of staff, he wrote helpful and cogent notes that let them read through confessions quickly.

"My notes simply tried to help my superiors grasp the substance of the confession," he says.

Duch is the master of confessions and, in his heart of hearts, proud of it. But he seems unwilling to admit the deceit that underlies his success. In other words, he doesn't see that all those confessions he compiled were born of terror and torture. He doesn't see how the climate of fear and the use of torture make his accomplishments deceitful and spineless.

At one point during the pretrial investigation, Duch seemed almost lucid when he told investigators: "I don't see what purpose confessions such as Tiv Ol's could have served."

"Tiv Ol was a teacher of literature, yet his confession had no purpose. What does that mean?"

"Do you think each confession served a purpose? We wanted truthful confessions. But they weren't confessing the truth."

"True or not true—what does that mean?"

"I don't know. That's the problem with beating people up and using violence."

Duch explained that he wanted to check with field agents whether confessions were true or not, but that Brother Number Two, Nuon

Chea, told him, "Comrade, you must think of the truth of the proletariat."

"What did he mean by that?" asked the investigators.

"I still don't understand it. Nuon Chea is someone who doesn't like to explain. The main thing was that the proletarian class emerge victorious, no matter what. He wasn't thinking in terms of justice or injustice."

Once the trial starts, Duch proves more reluctant to accept any fact that calls into question the performance of his duties:

> Son Sen informed us that they had found a CIA agent in Sector
> 32. He asked me why we hadn't found any CIA agents at S-21. I was
> speechless. From then on, we were required to find CIA agents. I spread
> the word and suddenly the confessions contained plenty of mentions of
> CIA agents. My job was to point them in the right direction.

"Did you really think there were so many KGB and CIA agents?" asks Judge Cartwright.

"They were probably forced to say that," says Duch hesitantly, as though saying it will send him over the edge.

"Were all the prisoners permanently subjected to a climate of absolute terror?"

"Quite frankly, yes. Definitely."

**SOME OF THE CONSPIRACIES** that emerged from the confessions were staggeringly coarse. For example, one nineteen-year-old girl, so terrorized by Prak Khan that she soiled herself, was convicted of sabotage for having defecated in the soup served in the Khmer Rouge command's hospital, as well as in its operation unit. A number of fantastically implausible plots were discovered, such as one about tunnels burrowed into the flooded bowels of the alluvial plain beneath Phnom Penh, where hundreds, even thousands, of Vietnamese soldiers were supposedly holed up.

"Was the point of the violence to extract truthful confessions or ones that conformed to what you wanted to hear?" asks Judge Lavergne.

> *I never believed that the confessions told the truth. Forty percent at best were true. As for the denunciations, only 20 percent of those were true. There was no scientific way of monitoring the confessions. There was no scientific method to ensure that they were true.*

The former math teacher's statistics fluctuate. He has previously declared that, at best, only 20 percent of the confessions reflected the truth, and 10 percent of the denunciations. To this day, despite knowing how specious they are, how much they are travesties of genuine confessions, Duch still needs to assess their value. As long as they retain some fraction of truth, it doesn't matter whether it is 40 percent or 20 percent or 10 percent. But he does admit the following: "I never believed that they represented the truth. Even the Party's Standing Committee didn't completely believe them. The point of it all was to get rid of people who were in the way."

Denouncing accomplices was crucial; Chandler reminds us that "you can't conspire by yourself." A conspiracy needs coconspirators. Every secret police in every Communist dictatorship—and in non-Communist ones, too—has compelled people to denounce others; everyone drew up imaginary lists. In this regard, S-21 broke no new ground.

The court openly hates the very idea of denunciation. Given that at S-21 thousands were tortured and mercilessly killed, the court vehemently rejects the validity of the denunciations obtained there. But in other circumstances, the international legal establishment can be more accommodating. Mandatory denunciation (though obtained without torture) is a crucial element in many confessions made before international tribunals and, in these circumstances, lawyers find that their consciences remain quite untroubled by it. On the contrary, they actively encourage it. A defendant who pleads guilty to a

UN tribunal is told to denounce his accomplices if he wants to win over the prosecutor and earn the judges' leniency. He isn't forced to name names under torture, of course, but if he wants to make the most of his guilty plea and obtain a lighter sentence, then he has no real choice but to comply. Rwanda's community courts, known as Gacaca courts, which have been so misguidedly praised over the past ten years, feed off of mass denunciations. Though they don't torture people, snitching is inextricably linked to confessions in Gacaca courts. The result is an all-consuming, rampant, and poisonous judicial operation that has produced more than a million suspects. Throughout Rwanda, the pressure to name one's accomplices has given rise to slander so great it would not be out of place in the archives of S-21. "Denunciation is another form of lying," François Bizot, a survivor of imprisonment by the Khmer Rouge, says in court. International justice, it seems, only hates lying in certain circumstances.

**EXTERMINATION CAMPAIGNS ALL HAVE** one absurdly equitable moment: they always end up devouring their own members. The concentric circles of those condemned to die contract with each round of killing, and new, increasingly restrictive criteria are established until the killing machine holds in its sights its own most fearful servants and greatest champions. At S-21, some prisoners denounced Son Sen—Duch's "master," the founder of S-21 and head of the Khmer Rouge security apparatus—saying he was working for the Vietnamese. Some even named the powerful and much-feared leader Ta Mok as one of those cancerous enemies to be excised. One prisoner said a woman called Khieu Ponnary was a CIA agent. In the margin, Duch wrote, "Whose wife is she?"

Khieu Ponnary was Pol Pot's wife.

Duch himself was implicated in the confessions of at least two prisoners: that of his former teacher who had introduced him to the Revolution, and that of his former boss, Brother Number Five, Vorn

Vet. A lawyer for the civil parties asks Duch why he suffered no consequences. Duch listens to the question, leaning back slightly in his chair. The teacher's accusations were weak, he says, because they were counterrevolutionary allegations dating from before the *maquis*. In other words, they were no longer relevant. "As for Vorn Vet, everyone knew that I had been in his debt. He wrote down my name. I changed nothing because people would've said that I opposed it. If I was to die, then so be it. I survived because I remained loyal. I survived because I was honest with them."

By the end of 1978, people increasingly close to Duch were being killed. Duch says that he felt "hopeless" and that he simply awaited his turn. He emptied his prison of its last inmates on January 3, 1979. After that, he says, he felt like he "was waiting to die."

"If Vietnam hadn't invaded Cambodia, Son Sen would probably have fallen," says David Chandler.

Had the Angkar eliminated Son Sen, Duch would have been next, along with his wife and children. His loyal deputies, Mam Nai and Pon, would have followed soon after, all of them dots on the same "line" of traitors, and instead of being on the docket today, Duch's name would be just another on the unfathomably long list of victims.

# CHAPTER 9

**E**UPHEMISM IS THE VERNACULAR OF MASS MURDER. UNDER Pol Pot, the regime didn't kill people, it "resolved" them, or else it "withdrew" them from their work or combat units. The regime "destroyed" or "smashed" people, depending on how you translate the Khmer farming term *komtech*: to reduce to a thousand pieces or, according to the translation adopted by the court, to smash.

"Resolve, smash, execute—they all mean the same thing. A person was executed and buried," says Duch.

Revolutions are plagued by such lies, duplicities, and dissimulations. The preamble to the Constitution of Democratic Kampuchea, for example, makes the following lofty, Promethean statement of purpose:

> Whereas the entire Kampuchean people and the entire Kampuchean Revolutionary Army desire an independent, unified, peaceful, neutral, non-aligned, sovereign Kampuchea, enjoying territorial integrity, a national society informed by genuine happiness, equality, justice, and democracy without rich or poor and without exploiters or exploited; a society in which all live harmoniously in great national solidarity and join forces to do manual labor together and increase production for the construction and defense of the country.

All this was just *"une façade,"* says Duch in French, intended to "disguise the dictatorship."

Anyone who wanted to be a Party member was required to adhere to ten conditions. These, too, were a sham, according to Duch, designed to keep people out. You had to have:

1. A solid revolutionary position within the Party line
2. A solid revolutionary position within the Party's proletarian ideology
3. A solid revolutionary position on internal Party solidarity and unity
4. A solid revolutionary position within the decisions, leadership, and work of the Party
5. A solid revolutionary position in maintaining the vigilance and secrecy of the Party and defending revolutionary forces
6. A solid revolutionary position in maintaining the independence, autonomy, self-reliance, and internal control of the Party
7. A solid revolutionary position in making and controlling revolutionary biographies and revolutionary self-criticism
8. A solid revolutionary position on class
9. A solid revolutionary position on clean life morals and on politics
10. The potential to self-edify and to be receptive to future leadership

In Duch's eyes, however, the regime's fundamental document remains the statutes of the Communist Party of Kampuchea.

> It was a document I feared so much that I had to constantly reread it. We had to measure our own philosophy against the statutes. I may not have followed Revolutionary Flag [the Party magazine] perfectly, but I adhered to its statutes to the letter. I worked very hard to uphold them, because they were the measure the regime used to decide whether you lived or died.

*The Party considers Marxism-Leninism to be the foundation of its vision and the guide for all of its actions.*

*Based on this principle, the Communist Party of Kampuchea absolutely resists and fights idealism, empiricism, bookish science, and reformism.*

*The Party fights in absolute terms the diseases of isolationism, authoritarianism, militarism, mandarinism, and bureaucratism. At the same time the Party is also opposed to lagging behind the masses.*

When I discussed the trial with young people born after S-21 was shut down, I found it telling and ironic to see the bewildered expression that this Marxist gibberish brought to their faces. Even the Cambodian court interpreters regularly had trouble understanding what was being said. The language of Marxism seems completely foreign to them now, jargon from another age. Words once familiar to everyone have become unfamiliar and hazy today.

Though Judge Lavergne experienced the Cold War, he was only nineteen when the Khmer Rouge fell from power, and he is wary of the regime's more baffling statements.

"What does the struggle against 'bookish science' mean?" he asks.

Duch tries to clarify as best he can. "The word 'dogmatism' might be more accurate," he says.

*We also used the term "dogmatism." It is based on the whole of Leninist theory. The former Soviet Union had a working class, but that wasn't the case in Cambodia. We had no industrial working class. So we had to control the peasant class, because if we had waited for a working class to emerge, then it would've meant that we were being dogmatic. Therefore, we did not exactly follow Marxism-Leninism.*

Cambodia's Communists intended to go faster and do things better than their Bolshevik and Maoist predecessors. They saw no need to wait for Cambodian capitalism to develop, nor even for a proletariat to form. They abolished money. They imposed radical agrarian reforms on the country. They decreed wildly excessive rice production quotas. Business was conducted by the whip and bayonet. And through it all, the Khmer Rouge leadership concerned itself more with eliminating those who stood in its way—or were simply born to the wrong parents in the wrong place—than with managing production. After all, the regime needed to find *someone* to blame for the catastrophic failure of its economic policies, and for the famine that ensued.

Expert witness Craig Etcheson points out that Khmer Rouge district leaders, in the reports they telegrammed to the Central Committee, devoted on average half a page to economic development, another half to production figures, and five pages to the hunt for the enemy within. This obsession with the enemy spread throughout the country and infected every level of authority. The exhausting and endless hunt for traitors sucked dry all of Cambodia's energy and talent—and S-21 was at its very heart.

> The Party must protect the revolution as much as possible from any action or trick conducted by an enemy in any direct, indirect, open, or secret way whose aim is to destroy the Party by all possible means. All Angkar Party organizations and every Party member must always be good, clean, and pure, in politics, mentality, and command, permanently, through a clean and pure biography, consecutively and constantly.

> *We had to be eternally vigilant in order to distinguish our friends from our enemies. We tried to prevent our superiors from being labeled left- or right-wing. Otherwise, it could've led to trouble for us. That was the core principle, and one of the conditions you had to meet before being admitted to the Party. We fought hard, but our enemies were legion. I didn't know that the Party's intention was to abolish civilization and humanity.*

"Did the party purge people who became or proved to be too extreme? Or was a good leader somebody, first and foremost, who could be controlled; that is, someone who could be counted on to do what he was asked as much as possible?" Judge Lavergne asked.

"The good leaders were those who did not act excessively, but did not make mistakes, and completed what had been assigned. So that was the main purpose: we had to make sure to follow the instructions for each assignment we received."

When a person was arrested, he or she necessarily became the enemy, and the enemy had to be systematically eliminated. By this brutal but effective logic of simply presuming suspects to be guilty, the Khmer Rouge ensured that its henchmen didn't flinch when extorting confessions from prisoners. To this day, Duch cannot give a straight answer to the question: Was it official policy of the Communist Party to kill and crush its people?

"Our aim was to be absolute and to defeat the enemy bit by bit. The language was slightly different, which may have led to some misunderstandings. Nowadays, lawyers call it extrajudicial killing, but at the time we called it class struggle."

Unlike the Party's official documents, its propaganda demonstrates a much clearer and more direct tone. The Party magazine *Revolutionary Flag* demanded "absolute measures," "zero tolerance," and "no hesitation":

> We have managed to wipe out 99 percent of our loathsome enemies hidden within. Now we must do the same thing throughout the entire country. Every sector must be investigated in this way. Every district must be investigated in this way. Every cooperative must be investigated in this way. The military, party officials, and the government must be examined in this way.
>
> Eliminate, eliminate, eliminate, again and again, ceaselessly, so that our Party forces are pure, so that our governing forces at every level and in every sphere are clean at all times.

# CHAPTER 10

**D**UCH'S IS THE FIRST INTERNATIONAL TRIAL CONCERNED WITH crimes committed in the name of Communism. International lawyers and human rights activists are scathing about so-called nationalist revolutions—those that openly pursue racist or xenophobic aims. No one has any trouble rejecting the movement for a Greater Serbia or denouncing Hutu Power. Yet many balk at the notion that, in a trial of the Khmer Rouge, Communism itself takes the stand as well.

When right-wing revolutions conflate purity with race, the violent resulting ideology is cause for alarm; yet when a left-wing revolution conflates notions of purity with class, it is somehow deemed appealing. Desiring a single race of men is a hateful project; desiring a single class of men (or two, or four), a good intent.

Before the trial started, I attended a public forum in Pailin, a Khmer Rouge stronghold during the twenty years of war that followed the overthrow of Pol Pot. The forum was set up to explain the tribunal's purpose and how it would operate. The international prosecutor said to the crowd:

> Here we believe that they wanted to take power to change their country, to make things better. But if you commit crimes, regardless of your belief, you should be prosecuted. The reasons are only relevant for the punishment. It's not an ideology that is on trial but people who committed crimes in the name of an ideology.

In other words, Pol Pot was evil, not Communism. The ideal is redeemed by good moral intent. From Stalin to Mao and from Kim Il-sung to Mengistu, Hoxha, and so many others, mass murder was committed with the same moral intent. And it would seem that, for some, good intentions are a mitigating factor.

It is in this vein that Nate Thayer, the famous journalist who obtained an exclusive interview with Pol Pot shortly before his death, gave his analysis on the Cambodian tragedy: "The Khmer Rouge wanted to modernize the country. Whatever their faults, the fact is that they weren't motivated by selfish reasons. Pol Pot wasn't corrupt. They were trying to lift Cambodia out of the feudalism and corruption that had brought the country to its knees."

Cambodia experts who would later work for the tribunal subscribed to the same idea. "Pol Pot may have had noble goals, but his methods were catastrophic," said Craig Etcheson, who later joined the prosecutor's office. "He accomplished the opposite of what he set out to do," claimed Stephen Heder, a key member first of the prosecutor's office and then that of the investigating judges.

"We loved the people and the nation, but practical mistakes were made," Pol Pot told Nate Thayer. "I must tell you: I did not join the struggle to murder my people," he said. At this point, he was a weakened man, cast out of any leadership role. A silence ensued. His eyes fluttered. He stared at the camera, a faint smile on his face. "Look at me: do I look evil? Not at all," he said, closing his eyes and waving his hand dismissively. "My conscience is clear."

In Pailin, the prosecutor, protected by two bodyguards, addressed the crowd. When he was finished, there stood in the middle of the meeting room a peasant, one of those humble people in whose name intellectuals directed the Revolution, and on whose behalf the tribunal says it is acting today. "You're convicting the smoke but you must see the *source* of the fire. I want to know *why* so many people were killed."

The village woman who spoke next only added to the embarrassment of those in power: "This tribunal does not bring reconciliation. Justice does not exist. We have elections, but power and wealth remain in the hands of people in power. If you are not rich, you don't get justice. Those who committed crimes are in power. They don't admit to it. They keep the power."

In court, Duch insists that ideology is important. And Nuon Chea, Brother Number Two and Duch's direct superior between 1977 and 1979, has said, "Ideology is truth. Truth comes from practicing ideology." Of course, the political philosophy to which Nuon Chea was referring led to some of the twentieth century's worst totalitarian regimes, and many intellectuals who embraced it then now squirm when confronted with its history. The lawyers at Duch's trial appear curiously sensitive to this, as though fearful of somehow being undone by it. In response, they champion a doctrine of personal responsibility and maintain a scrupulous—and convenient—stance of political neutrality. Lawyers like to say that determining *why* a crime occurred is not within the purview of a criminal court—yet it's the question on everyone's lips.

In the end, it's the Cambodian prosecutor who, at the start of Duch's trial, tries to answer the common-sense question posed by the peasants of Pailin.

> *For thirty years, 1.5 million victims of the Khmer Rouge have been demanding justice for their suffering. For thirty years, the survivors of Democratic Kampuchea have been waiting for accountability. For thirty years, a whole generation of Cambodians has been fighting to get answers about their families' fates. Today, at long last, this process begins and justice will be done. You are also called upon to determine why it happened, because history demands it. The purpose of courts such as this one must be to establish the truth, unflinchingly and without fear. The ultimate goal of the Khmer Rouge was the establishment of a "pure" Communist society unlike any seen before. Some then and perhaps still now argue that the Khmer Rouge came to power with the best of intentions and that something went terribly wrong. But that is simply not true. From the very beginning, the Khmer Rouge leadership was intent on ridding itself of its perceived enemies.*

Long before the Khmer Rouge, some twenty years before Democratic Kampuchea came into existence, when Duch was still a schoolboy, the philosopher Raymond Aron wrote a scrupulous analysis warning of the chimera of good intentions. He wrote in *The Opium of Intellectuals* (1957):

> The idolaters of history cause more and more intellectual and moral havoc, not because they are inspired by good or bad sentiments, but because they have wrong ideas. [ . . . ] The essence of a political regime is found not in the principles it proclaims, nor in the ideas it calls its own, but in the life that it gives its people.

In 2009, there are no anti-Communist activists using Duch's trial to substantiate their warnings of the dangers of Communism; this is a given. Rather, Duch's trial seems to interest mostly those old Marxists who have more or less kept the faith. They come primarily, it seems, to continue arguing among themselves. Revolution is a drug that lasts, with its promise of romance and excitement and its intellectual distance from the tedium of reform. Marxism is the opium of intellectuals, wrote Aron, just as Marx said that religion is the opium of the masses. It is Westerners who will write history in Phnom Penh, just as they did in Rwanda and Sierra Leone, and just as they did at the temples at Angkor. Some of them see in Duch's trial a tacit retelling of their own broken dreams.

**THE CAMBODIAN GOVERNMENT CLOSELY MONITORS** the tribunal. The state's three highest-ranking members are themselves former Khmer Rouge officials who fled the purges and escaped across the border to Vietnam in 1977. Other important ministers and many military officers are also former revolutionaries. Nothing illustrates the atmosphere of hushed tension and ambiguity hanging over the court better than the "Jarvis Affair" that breaks out in the middle of the trial.

Duch is in the dock talking about the policies of the Communist Party of Kampuchea when Helen Jarvis, who has been running the tribunal's Public Affairs section from the start, is named head of Victims' Support. Because Jarvis's longtime Communist affiliation is no secret, the announcement causes a bit of an outcry. But the tone becomes far more strident when a political manifesto surfaces, which was written as recently as 2006 and which Jarvis signed.

The text is titled "We Are Not Leaving the LPF!" The Leninist Party Faction (LPF) was a splinter group of the Democratic Socialist Perspective, a now-defunct Australian far-left political party. The faction and the political alliance to which it belonged was a world of infighting, a place where purges and plots abounded, where people accused one another of treason and of plotting to secede, of vindictive and petty-minded fights between "comrades," violations of Party discipline, collaboration with the enemy, and crushing insults. Its members still dreamed of being at the forefront of the Revolution and celebrated the one started under Hugo Chavez. The future of World Communism, they said, revolves around the Venezuela-Cuba axis. They respected "democratic centralism" and vilified the "international capitalist media."

The manifesto's signatories were unhappy with the line held by the party leadership, because it failed to promote "the Leninist strategy of building a revolutionary party made up of Marxist cadres— the key to advancing the socialist struggle." They went on to say that "We too are Marxists and believe that 'the end justifies the means.'" They adopted an ambiguous position: "For the means to be justifiable, the ends must also be held to account," but then state quite unambiguously: "In times of revolution and civil war, the most extreme measures will sometimes become necessary and justified. Against the bourgeoisie and their state agencies we don't respect their laws and their fake moral principles [sic]."

The text, which Duch wouldn't have found objectionable in the 1970s, had been signed by the person that the tribunal put in charge of looking after the victims of the Khmer Rouge's revolution. As Aron wrote: "Communist faith justifies all means, Communist hope forbids acceptance of the fact that there are many roads toward the Kingdom of God, Communist charity does not even allow its enemies the right to die an honorable death." He continues:

> The sublime end excuses the revolting means. Profoundly moralistic in regard to the present, the revolutionary is cynical in action. He protests against police brutality, the inhuman rhythm of industrial production, the severity of bourgeois courts, the execution of prisoners whose guilt has not been proved beyond doubt [ . . . ] But as soon as he decides to give his allegiance to a party which is as implacably hostile as he is himself to the established disorder, we find him forgiving, in the name of the Revolution, everything he has hitherto relentlessly denounced. The revolutionary myth bridges the gap between moral intransigence and terrorism.

Some civil parties write of their distress at the Jarvis affair. The court's reaction is to ride out the storm. A UN spokesperson tells the media: "Every staff member has a right to personal political views."

Helen Jarvis herself says little. She feels wounded by the victims' challenge. "I have spent ten years working to establish this tribunal," she reminds her opponents. She has written a book about it.

In any event, the victims of Cambodia's Communist dictatorship never organized themselves into an advocacy group. They carry very little weight.

The storm passes. Ideas and ideology, we're told, aren't relevant. Jarvis keeps her job.

But unlike Jarvis, who can dodge the media, Duch cannot dodge Judge Lavergne's questions in the courtroom.

"Did you believe that the end mattered, not the means? Was that the way you saw it at the time?"

"Exactly," replies the former Leninist.

# CHAPTER 11

ONCE THE MIDDLE-RANKING OFFICIALS PRAK KHAN, SUOR THI, and Him Huy have given their testimony, it is the turn of the lower ranks.

One former guard wears rectangular glasses with a thin, dark frame. He parts his well-cut black hair low on one side, his hair so thick that the part is barely visible. It is so neatly trimmed that he looks like he's wearing a wig. He is fifty-one years old, with the demeanor of a well-behaved schoolboy. Thirty-five years earlier, he served as a guard and messenger at S-21, the bottom of the ladder. He sits straight and still in his seat, and positions his forearms confidently on the armrests. He looks like he's sitting in the electric chair. When he speaks, he purses his lips at the end of each sentence, which makes him look like a prude who's just heard a swear word. He ends each answer by clenching his jaw, swallowing, and lengthening his neck like a turtle.

The former guard's brother was arrested and killed at S-21, and the witness says he was terrified that he would be, too. He says he owes his life to Him Huy, the head guard. But he levels serious accusations against Duch. Before the trial, he told investigating judges that when a prisoner's answers weren't clear, Duch would say to them, "Are you going to talk?" then hit them a couple of times before adding, "You'll know soon enough." The former guard swore it was true. He also said that Duch came to the prison almost daily.

But in court, he plays all this down, saying that he had been "a bit excessive" in his earlier testimony. He describes to the court the severe interrogation as he remembers it.

"When I came back from lunch, I saw Duch in a villa next to a wooden building. I saw that and it's the truth," he says, before swallowing and extending his neck.

"Did you see Duch beat up the prisoner?"

"Duch used a rattan cane. He didn't beat him too much before I left."

"You saw this with your own eyes?"

"I saw this with my own eyes because I was guarding the two-story building and that's the truth."

"Did you see him torture other prisoners?"

"No."

Wearing a pale lilac shirt, untucked, Duch is dressed like a member of today's Khmer ruling class. He says he feels sorry for the witness, whose testimony is generally accurate—except, of course, for the bit implicating Duch directly in the interrogations. "My crime was indoctrinating personnel. That was my crime against those who weren't arrested by the Angkar. The interrogators resorted to torture because they had to. I don't deny that. But the guards had to do their job, not conduct interrogations."

Duch recognizes before the court that the witness, now a farmer, was a combatant who suffered greatly. He says he shares this man's suffering and offers him his condolences. He sits down and carefully puts into a plastic sleeve the documents he used during his statement. He looks at the former guard before he leaves the room. Duch slides the plastic sleeve into a big red binder.

Another rice farmer takes the stand. He's wearing the gray suit jacket in which a succession of witnesses has now appeared, each of them swimming in it. Whenever he doesn't understand a question, he grins. A French advisor to Prince Sihanouk wrote that the Khmer smile

[C]onceals brilliantly the true feelings of men. Throughout the Far East, the smile is the polite mask from behind which people watch one another, congratulate one another or fight one another. But in Cambodia, this mask is more often than not one of pleasant-enough and ambiguous indifference that people hold up between themselves and others. One should never mistake a smile for an invitation to start a conversation. On the contrary, a smile signals the worry and embarrassment provoked by an outsider. The smile, in Cambodia, indicates that one has neither the intention to answer indiscreet questions nor to ask them.*

The farmer leans toward the microphone, his eyes twinkling. He has the easy good cheer of one so unaccustomed to the pomp and ostentation of the rich and powerful that they cannot intimidate him. He was fifteen years old when he was sent first to the S-24 reeducation camp, then to S-21, where he worked as a guard in Building B. He entertains the court with his naive, transparent answers and the smiles with which he punctuates them; a murmur of appreciation rises in the public gallery. The witness knows little and remembers even less. He neither saw nor met Duch, and he has forgotten what he told the investigating judges only last year. He is illiterate—or, at least, he was at the time of the crimes—yet the court asks him to confirm whether the prison rules and regulations were posted in each cell. He initially told investigators that he had witnessed numerous rapes. Now he says he saw none.

Duch explains that the witness was exactly the type of person he had sought to recruit: a young, uneducated, mentally and politically pliable member of the "base people."† In court, the farmer comes across as a useless witness.

Criminal investigations work in mysterious ways: prestigious international courts do not keep count of the sensational statements made in confidence during the investigation and then publicly disavowed at trial. If, in the absence of documentary evidence, a court

---

\* Charles Meyer, *Derrière le sourire khmer* (Paris: Plon, 1971)
† "Base people" is the original revolutionary term used by the Khmer Rouges to refer to the rank-and-file, or common people.

must issue a judgment based solely on testimony, then that judgment will often be little more than an act of faith. The more trials you follow, the more you disbelieve everyone: witnesses, the police, judges, prosecutors, defense lawyers, and victims.

The abundance of documentary evidence at Duch's trial makes it tempting to fall back on the principle set by the defendant himself: where there is no documentary evidence, there is reasonable doubt; and if there is doubt, you must not convict. The abundance of archival material takes pressure off of everyone's conscience: there's no need to believe in justice in order to form an opinion about it.

**WHEN A CERTAIN LY HOR** takes the stand and claims to be a survivor of S-21, Duch immediately makes it clear that he doesn't believe him.

Once a Khmer Rouge soldier, Ly Hor tells the court how, after his arrest, he was transferred from S-21 to S-24 (which never happened); that the roof over his cell was of corrugated iron (which the one at S-21 wasn't); that he was taken out to wash every three days (even though it's common knowledge that the prisoners were hosed down in the cells); that he was served rice (even though no prisoner at S-21 was allowed it); and so on.

Duch smiles, leans forward over his desk, intrigued by the poor wretch testifying. The victim's lawyer, tricked by the local NGO he works for, endures the worst day of his short career. Judge Cartwright upbraids counsel for the civil parties for their lack of preparation, and laments the consequences for those put on the stand. Judge Lavergne is furious about the "utter vagueness" of the documents relating to someone called Hor, from a certain "Bureau 43-44," which are supposed to prove that the witness was indeed a prisoner at S-21. The lawyer has no answers. Nor does the prosecutor.

But Duch does.

Off the top of his head, he remembers that in one of the case documents—"Document B-57, Appendix 003"—Bureau 44 appears to be that of the Army Division 703. As for Bureau 43, he adds, no documentation exists. He hazards a guess, but is careful to say that there's no proof. The court adjourns. After the break, the prosecutor returns with new documents, ones that Duch has never seen. He asks to see them. He flips through them, then embarks on a fascinating reconstruction of the possible path of one Ear Hor, the S-21 prisoner as whom Ly Hor has tried to pass himself off.

With his arm folded behind his hip, Duch looks almost professorial as he carries out a thorough analysis of the archives. He has already found in them the prisoner's date of execution. He cites the legal classifications by heart. He notes that three years separate the birth dates of Ly Hor and Ear Hor. And you only have to compare Ly Hor's document filing for civil action with Ear Hor's written confession, says Duch, for the fabrication to become blindingly obvious. "I can see that the handwriting is 50 percent different. Comrade Ear Hor and Mr. Ly Hor are therefore two different people. Finally, Ear Hor is dead and I will do nothing which may offend his soul."

Duch the super cop, whose talents had long ago been singled out by his Khmer Rouge masters, has just given a brilliant performance at his own trial. The court has wasted the day on one witness's fabrications. With just a few words, Duch has generated a consensus and in the process done himself the favor of coming across as more respectful of the victims than those who speak on their behalf.

Duch has an excellent grasp of the documentation and can find his way through it with a mathematician's precision. Everyone is amazed when he cites without notes the classification reference for such and such a document, or the page to which he is referring. Duch compiles, compares, checks, and memorizes the relevant documents. It's true that, when working with the various lists from S-21 (some of which overlap), Duch reaches conclusions that any diligent and painstaking accountant could have reached. But it's at this point that Duch adds something that only he can bring: an intimate knowledge of the institution that generated all of this data.

Duch's trial is at this point the only one taking place at the Extraordinary Chambers in the Courts of Cambodia (ECCC). Just four more Khmer Rouge leaders are scheduled to be tried after him. This means that many victims won't share in this rare moment of justice, for the simple, unbearable reason that their prison or their cooperative or their dyke or their canal wasn't one of the sites selected during the investigations. Courts such as the ECCC, that render symbolic justice, are faced with the bitter task of choosing which victims will have their day in court and which won't. For Ly Hor, the lure proved irresistible.

"When I heard about Duch's trial, I became determined to be part of it. I suffered so much under the Khmer Rouge," he says.

Another alleged survivor of S-21 takes the stand. He is a handsome man with hair cut short, a round face, and laughing eyes. His eyebrows are slightly arched, and his lips are very pronounced and delicately outlined, not unlike those of the great Khmer king Jayavarman VII, whose face watches over his people from the four corners of the famous Bayon Temple. The man on the stand has a calm voice and keeps his eyes fixed dead ahead.

He remembers being sent to S-21 in 1976. He remembers being given a little fish in his food ration, being let out of the cell to wash, and being sent out to work in the vegetable garden. Again, you would think that anyone who knows how S-21 worked would have immediately unmasked this witness. But the prospect of parading before the court a survivor unknown to anyone for thirty years proves too great a temptation for the NGOs, and their discernment fails them.

DC-Cam, the Document Center of Cambodia, was founded in the mid-1990s by Yale University. Within a decade, it had built Phnom Penh's largest archive on the history of the Khmer Rouge. All of the S-21 archives are stored there, and when the tribunal was established in 2006, it proved an indispensable resource. But it was DC-Cam, perhaps swept up in the frenzy whirling around the tribunal, that "found" the daring substitute Ly Hor. Not to be outdone, the NGO Lawyers Without Borders produces its own improbable survivor.

Faced with such fecklessness, Duch hardly needs lawyers. His opponents turn out to be his best defense. Even worse, they embarrass some of the victims. Judge Cartwright is furious, her face locked in a scowl. Judge Lavergne looks at his fingers, then around the room, careful to avoid meeting the eyes of the reserve judge, though she's obviously as stunned as he is. The spectators who remain in the public gallery have—wisely—gotten unruly, and present a more genial scene than the one taking place in the courtroom: some villagers have noticed that they can see themselves in the background on the screens. They smile and chuckle and joke around. The air smells of Tiger Balm. There's a palpable buzz, for this is no ordinary session: it's a big day in court! Though there's little doubt that the poor fool on the stand suffered physically and psychologically at the hands of the Khmer Rouge, it wasn't at S-21, and no one bothers listening to him any longer. He may not have been locked up in Duch's prison, but he is nevertheless one of Cambodia's millions of victims. Sometimes, he tells the court, pus still oozes from his left ear.

Next, an ex–Khmer Rouge soldier limps into the courtroom. He comes across as a humble and gentle man, despite his eyes sunken into their sockets, his hollow cheeks, and his sharp, angular face. The soldier helped to clear Phnom Penh of its inhabitants in April 1975. The following year, he was arrested during one of the many purges in the Northern Zone. He was reinstated as a radio operator, then rearrested in 1978. He gives a flawless description of his detention at S-21. He wasn't photographed upon arrival, he says, but all his other details match. Unlike those of the fake survivors, his story contains no credibility gaps. That is, not until he describes being taken away to be executed. "It was probably the night of January 6, 1979," he says, because he could hear gunfire. He was blindfolded and made to kneel beside a ditch. He told himself that his hour had come. Someone hit him in the ribs and he tumbled unconscious into the hole. Then, at around two in the morning, he came to. He didn't know where he was. He felt dizzy, but managed to free himself from the rope binding him. He could smell blood. He climbed out of the ditch. There were no guards in the vicinity. It was only later, he said, that he realized that the place was Choeung Ek.

He walked for an hour before lying down on a tree trunk. He was starving and tried to chew on a banana stem. He found his way to the river, where he put a trunk in the water, lay down on it, and pushed off the bank. He drifted downstream for two or three days before reaching the place known in Phnom Penh as the Japanese Bridge. Members of the army that had just driven out the Khmer Rouge found and rescued him. He remembers that there was still intense bombing going on. He was ill but safe.

The witness speaks in even tones for almost an hour without pause, barely moving, his eyes riveted to the floor. Thirty years after the fact, like some magician pulling a rabbit from his top hat, the human rights association ADHOC (along with Lawyers Without Borders, who represent the witness in court) has just presented to the world the only survivor of the killing fields of Choeung Ek, and the sole survivor whose violent interrogation Duch would have personally attended.

But there's a problem, and it lies in all the other statements the miracle witness made before Duch's trial: they differ substantially from what he tells the court. When he is confronted with this, the ex-soldier's memory suddenly fails him, even though he remembered the tiniest details when telling his tale to the court. A murmur rises in the gallery. The public, made up mostly of peasants, has a sharp ear for nonsense.

"This is utterly different from what you have previously told us," snaps the presiding judge, determined to expose this miraculous survivor.

The witness, suddenly nervous, begins to blink rapidly. He confesses that he had visited S-21 in 2008, during a trip organized by ADHOC, with the purpose of looking for the "biographies" of his cousin and his cousin's wife. That's when he first learned about Choeung Ek—the place that he claims never to have revisited since he dodged death there all those years ago.

Now it is the judges who want to try a little police work. All the documentary evidence—the lists of those arrested and executed, the thousands of confessions, the photographs, the biographies—constitutes the vast terrain on which they play a kind of legal scavenger hunt. Bit by bit, each player has been drawn into Duch's well-oiled machine. He has shown them the way, has demonstrated to them the efficiency of his system. Everyone knows how to use it now; everyone knows how to cross-check facts. Digital technology makes it even easier, and the players find the game irresistible. Another false "survivor" is unmasked. Though she's loath to do it, his lawyer must beat a hasty retreat. She finds others to blame: "The human rights organizations collect the testimonies. The work is done by young, untrained investigators. They are amateurs using whatever means they have available."

# CHAPTER 12

**A**S SOON AS THE VIETNAMESE FORCES THAT TOPPLED POL POT discovered S-21, they made a propaganda film about it. Shot in early January 1979, the film shows the blackened and bloated corpse of one of the last victims of the regime's secret police—a Khmer Rouge soldier, according to Duch. The body is lying on an iron bed. Behind the bed is a small desk with a typewriter on it; then a shot of the courtyard, where crows are landing by a decomposing body. The army reporter's voice announces, "We found these children in an office." Three children appear in the frame: one is wearing a white shirt and a cap, two are wearing the black uniform of the Khmer Rouge. One gives the camera a hard, challenging look. The camera zooms in for a close-up and you notice that "he" is wearing earrings. Another looks up toward the ceiling. The camera pans right and we see a half-naked baby lying on a mat. The commentator tells us that the children are "paralyzed with fear and hunger."

Norng Champhal is the oldest of the children found at S-21. He was nine years old at the time. Seven months later, he testified to the people's court set up by Cambodia's new government in order to judge the crimes of the "Pol Pot–Ieng Sary Clique" and to assert its own legitimacy in the process. The forces that deposed Pol Pot's Communist regime were themselves Communist. The Soviet Union and China fought their proxy war in Vietnam and Cambodia, and things can get confusing when Communists put other Communists on trial. The Vietnamese-sponsored court denounced the Khmer Rouge regime as having been backed by "reactionary forces" in Beijing. Cambodia's Maoists were reviled as "counterrevolutionaries," "imperialist lackeys," even "fascists."

Several witnesses called to testify before the court ended their depositions with a jarring cry of "Long live the revolutionary forces!" In a few short months, little Champhal had already had enough time to learn to use the term "Polpotists." In his testimony at the time, he described in extraordinary detail what he saw and experienced at S-21:

> *Each time Polpotists got angry, they beat us mercilessly. They hit us on the head. They kicked us in the back when we did not go quickly. Once, after we heard gunshots, my brother and I hid behind a heap of clothes taken from the prisoners. At that moment, I saw they were killing a boy a little older than me. He was bashed against a tree beside the kitchen. I did not know where they threw the body of the boy. [ . . . ] While I was in the prison, I saw the most atrocious tortures by the Polpotists against prisoners. They burned an iron stick and used it to perforate the noses of the prisoners. The women prisoners were plunged into water tanks. Some days before their departure, they showed me a photo of my disemboweled mother. [ . . . ] Once, after lunch, I saw five Polpotists taking a prisoner wearing white knee-breeches and a blue shirt to the gallows. After knotting [the rope around] his neck, they pulled up the other end of the rope in a way that the poor prisoner rose in the air. Then they loosened the rope and let the prisoner fall down from a high gallows. The victim suffered this sort of torture for the second time, before his body was dragged to a cell beside the electroshock room. After some time, they brought out another prisoner, who had on only knee-breeches. They killed him in the same manner. After his death, I saw his tongue [fall] out of his mouth. And then they led the third, who went slowly, because his hands were busy holding up his unbraced knee-breeches. They beat and kicked him in the back to make him go quickly. He was then hung up in the air. As his knee-breeches slipped down to his feet, the Polpotists burst into gleeful laughter.*

---

\* "Testimony of Surviving Prisoners, Investigation Report, People's Revolutionary Tribunal Held in Phnom Penh of the Trial of the Genocide Crime of the Pol Pot–Ieng Sary Clique," documents collected by "a group of Cambodian jurists" (August 1979), pp. 134–35

When the liberating forces found them in S-21, Champhal and the other children were taken to an orphanage. Champhal was then adopted by Keo Chanda, an important member of the new ruling power, which comprised Khmer Communists opposed to Pol Pot who had returned in Vietnamese tanks. By coincidence, it turned out that Keo Chanda, the minister of information and propaganda at the time, was also president of the People's Revolutionary Court, in which young Norng Champhal had testified. The little boy was introduced to Hun Sen, the young and ambitious minister of foreign affairs, who went on to become prime minister in 1985 and who has held on to the post ever since. Norng Champhal also met the real survivors of S-21, including Vann Nath. We don't really know what became of the child survivor after Keo Chanda died at the end of the 1980s. Everyone who has studied S-21 knows of his existence; some met him in the early 2000s. Yet when the ECCC was established a few years later, its investigators showed little interest in him. Indeed, it seemed as though everyone had forgotten about little Norng Champhal, whose testimony as a ten-year-old had been so sensational at the show trial of 1979.

Everywhere from Kigali to Phnom Penh, from Sarajevo to Baghdad, you will find men and women who have devoted their lives to bringing to justice those responsible for mass murders. Without their drive and commitment, they would never have the energy to carry out their work as activists, researchers, curators of memory, victim advocates and unwavering enemies of state violence. One can only admire the steadfast, single-minded, and pure-hearted toil undertaken by these enemies of impunity. And without the determination—the *absolute* faith, as Duch would put it—with which they stand guard against the rest of the world's propensity to forget, and against the impure pragmatism of the powerful, there would be no archived record of these crimes, nor any hope of retribution.

That such tribunals exist at all is thanks in part to the efforts of people who, for reasons known only to themselves, have dedicated their lives to pursuing justice. And when the trials for which they have labored in obscurity for years or even decades at last take place, these people often find themselves blinded by the glare of the media and public opinion, and unsettled by all the money that suddenly appears. It can prove to be a difficult moment. When a court is established, it is recognition of their work, but it also threatens to sideline them. For the international lawyers who swarm to the court in order to dispense justice in a matter about which they know nothing, these men and women are vital experts and mediators. The lawyers immediately reach out to them. At first, these specialists are pampered, praised, and heeded. Then, as the members of the court gain in prestige and become increasingly autonomous, things begin to turn, and the relationship between those who have dedicated their lives to the cause and those for whom it is merely a career move ends in disenchantment, embitterment, and divorce. In the end, such courts almost invariably disappoint those who have invested their heart and soul to bringing the truth to light.

Among themselves, activists can be a hard-hearted bunch. Theirs is a world riven by bitter quarrels, scandal-mongering, jealous slander, and vindictive scheming. Taken one by one, they are each remarkable individuals; but when they are forced to share the crushing memories, they become carried away with the passion of old lovers. The world of guardians of memory and righters of wrongs, whether they're from Cambodia or elsewhere, is neither virtuous nor particularly kind. To the activists themselves, it can sometimes seem overrun by enemies.

In Phnom Penh, the collaboration between DC-Cam and the judiciary began cordially enough. But as the court started to make its presence felt, the NGO feared losing its preeminent position, and conflicts arose between the two institutions. Two years later, by the time the tribunal was ready to try Duch, its relationship with DC-Cam had degenerated into one of thinly veiled hostility.

Just before the trial opened, DC-Cam announced with much fanfare that it had found a child survivor from S-21. Norng Champhal made his appearance. Yet, contrary to what some were claiming, he had not just been "found."

A press conference was convened; the survivor appeared, now thirty-nine years old. More drama: Norng Champhal had missed by a mere two days the court's application deadline for inclusion in the trial as a civil plaintiff. DC-Cam asked the court to take into consideration his exceptional and moving case. The prosecutor stepped up to the plate and called Norng Champhal's testimony "essential," even though he was only a child at the time.

Three decades earlier, Cambodia's new, post–Khmer Rouge rulers had used the boy as a propaganda tool. Now, in the media uproar that flared up before Duch's trial, Norng Champhal again found himself a pawn in a game the stakes of which were far above his head.

While all this was going on, DC-Cam also claimed that it had obtained a never-released Vietnamese film shot at S-21. The prosecutor's office wasn't told about it until four days before the trial's preliminary hearing. The tussle between special-interest groups and egos intensified—all on the victims' behalf, of course. This was not a new phenomenon, nothing is new. But the demagoguery and hypocrisy in the air that day left behind a lasting sense of unease.

The judges didn't cave to the pressure, and Norng Champhal wasn't permitted to assert his victim's rights in court. He is now, however, called as a witness.

Norng Champhal has kept an almost adolescent voice. He has the brown skin of rural Khmers. He keeps his eyes glued to the ground. Duch slips on his spectacles, reads a document, looks up, removes his glasses, and takes a long look at the witness. The defendant is leaning back in his chair, in that relaxed posture he sometimes adopts when he's not particularly interested in the proceedings. Norng Champhal is talking about his mother being photographed upon entering the prison. He starts crying. Duch sits up straight.

Norng Champhal's 2009 deposition contains none of the terrible atrocities that he described in 1979. He describes the painter's studio, the wretched gruel, mosquitoes, gunshots, the hiding place beneath the pile of clothes, his liberation by the Vietnamese troops, being sent to the orphanage. He tells how he saw mutilated corpses abandoned on iron bed-frames and how he fled in fear. There's nothing that isn't already known in his deposition; nothing of substance, from the court's point of view, except for the miraculous story of a handful of children who somehow ended up at S-21 on January 1, 1979, two days before the prison was entirely emptied of inmates. Some prison staff took care of them, more or less. Incredibly, the children survived the final annihilation before being rescued six days later, when the city was liberated. In court, nobody challenges Norng Champhal's 1979 testimony. Nor do any of the judges, prosecutors, or lawyers call into question his more-recent interview with DC-Cam in which he claimed, erroneously, to have been interned in S-21 for "three or four months" before the liberation.

It's probably better this way.

During the adjournment, the defense team holds an agitated meeting. Duch seems especially nervous, his movements quick and sharp. He keeps his fingers pressed together, as though his former life as a soldier has left him forever a quick snap of the arm away from a salute. He gesticulates emphatically. His mangled left hand seems stuck to him like a flipper, which only underscores his agitation. The black robes flock around him; he gives his instructions.

Duch first thinks that Champhal's father *hadn't* been killed at S-21. Then, after consulting the documentation, he concedes that Champhal's father was killed at the prison, after all. Then, a few days after Norng Champhal has given his testimony and his mother's "bi-ography" is finally produced in court, Duch admits to the truth contained in the archives.

Every now and then, while giving testimony, Norng Champhal gives a furtive sideways glance. He regularly weeps, his tears symptoms of a distant and irretrievable trauma pervading a life that others have partially rebuilt for him. What part of his suffering comes from being an orphan? Which of his woes bears testament to the troubles that unscrupulous adults sowed in him? Eventually, a stained veil will be drawn over this bitter episode of Duch's trial. But what it has made clear is that the officials at the People's Court of 1979 were not the only ones to stage the kind of tasteless scenes usually associated with show trials.

# CHAPTER 13

T O BE A TORTURER USUALLY MEANS TO BE ON THE "RIGHT" side of a dictatorship. Yet even those who stood to gain from the Khmer Rouge's atrocities found that quality of life is a relative thing. No one in Democratic Kampuchea was allowed to visit their parents. No one retained custody over their own children. Children under six were entrusted to older women. Those over six lived according to the collectivist mold, in which they were supposed to spend the rest of their lives.

Enjoying oneself was out of the question. Falling in love was considered bourgeois. Married life was limited to seeing each other one day out of every ten. Couples avoided spending quality time together. All marriages had to have the Party's blessing. Sex outside of marriage was a criminal offense. Communism adhered to a strict puritanical ideology.

Duch weighed forty-nine kilos at the time. He was entitled to a small bowl of rice and two dishes a day, and he points out that his was a privileged diet. One former staff member describes how he was so hungry he resorted to eating the fake medicine that the painter Bou Meng described as rabbit pellets. Every three or four months, Duch went to a place that served Chinese beer. Going more often would have aroused suspicion.

To ensure better control over staff and to keep the security perimeter around S-21 completely sealed, the facility was run as a family affair. The five members of the female interrogation team, for example, were the wives of male S-21 interrogators. Their team leader was married to Hor, S-21's second-in-command.

The wife of another staff member worked in the kitchen. This system made internal purges easier to carry out. Whenever a staff member fell from grace, his wife and children could easily be "resolved."

Duch met his wife in 1974. They married in December 1975, four months after the creation of S-21.

"I was a cadre of Democratic Kampuchea. The woman I married was a member of the Communist Party of Kampuchea. I wasn't allowed to marry any 'April Seventeeners,'" he says, referring to those doomed souls who only came under the Khmer Rouge's authority on the day of the victory of the Revolution, April 17, 1975.

Their first child was born in April 1977 and a second in December 1978, three weeks before the prison closed and Duch's family fled Phnom Penh.

> *When the children were born, they were children of the Angkar.*
> *Let me be clear: that doesn't mean that the children of the Angkar*
> *had to spy on and denounce their parents; it meant that they had*
> *to adhere to the Communist ideology and be loyal to the Angkar.*
> *All revolutionaries wished for their children to love and join the*
> *Revolution. S-21 was part of the Revolution. I didn't think this*
> *meant that my children would become police officers like their father,*
> *but we did want them to follow the revolutionary line. The Party*
> *considered the children of cadres or peasants property of the Angkar.*
> *The Angkar was their parent. Whatever their status, they were*
> *children of the Angkar. And at the time I saw it that way, too.*

In the eyes of the Khmer Rouge, the only time a family could be brought together in the traditional sense was when its members were to be executed. If parents were killed, then their children had to be, too. Being a child of the Angkar guaranteed no other privilege than that of once again becoming the child of your parents on the day of their damnation. Such should have been the fate meant for little Norng Champhal.

Judge Cartwright asks Duch if the estimate of children killed at S-21—1 percent of the total number of victims—is correct. Duch replies that an archival document attests how 160 children were sent to the execution field *in a single day.*

"That's more than 1 percent," says the mathematician.

"I think you're correct," agrees the judge, her voice tight.

By the time Duch's first child was born, he had given ten years of his life to the Party. In court, he tries to downplay the fact that he enjoyed the rare privilege of living with his wife and baby more or less as a family. His children were too young for school at the time, and in any case, all the schools had been closed. The children of cadres had no more access to education than did other children. The party made sure everyone was equally ignorant.

As a teacher, Kaing Guek Eav had passionately believed in universal education. He had devoted himself to his students, strongly encouraging them in their pursuit of knowledge. So when he is asked in court what he thought of the sudden abolition of the education system, he can't bring himself to give a straight answer. Some contradictions are so great he can only come at them circuitously. "The education required by the Party was that everyone be loyal to it, perfectly loyal. We had to devote ourselves with absolute determination to the Party. That was the requirement to get a job."

"Did you have an opinion about that policy?"

"Although I saw it as a challenge that would be difficult to achieve, I had no choice but to apply it. There was no way I could question it."

"Why not teach?"

> *Education in a Communist regime is different. The Communists taught us that truly loving the people meant bringing them the dictatorship of the proletariat. We weren't allowed to teach things like logic or the Universal Declaration of Human Rights. And if we didn't follow this, they cut off our heads.*

**DUCH BECAME A FATHER** while he was running S-21 and sending hundreds of children to their deaths. No one understands this. Even Duch struggles to understand it. The execution of children and babies was a part of his administration with which he wasn't overly concerned. It's also one of the crimes with which his conscience struggles.

> *I can't seem to grasp this point in detail. This image of children smashed against a tree—I believe that happened. It was the work of my subordinates. But babies being thrown from the second floor? I don't believe that. In front of the prisoners? It couldn't have happened. Nonetheless, I admit that I had come around to the revolutionary view that the children of executed prisoners might seek revenge later. Therefore, I am criminally responsible for the murder of those children.*

"When you left home to go to your office, did you ever think about what distinguished your family life from the lives of the prisoners?"

"Your Honor, it had become commonplace. It was the same thing at M-13—we had to implement policy," he says, referring to the camp he ran before S-21.

"I asked you if you never thought of the contrast between your family life and the lives of the families at S-21 . . ."

"Yes, I thought about it. I was already thinking about it at M-13. But we were brainwashed with an ideology that made a clear distinction between the terms 'enemy' and 'friend.' I really wanted to have children. If I didn't die, I could build a family. And if we had to die, well then, we would die."

Other than psychologists, few people understand this double life. The human capacity to keep different areas, different activities, and different thoughts of one's life separate is a well-known psychological mechanism. It's called compartmentalization.

"Duch's inner world is separated into watertight bastions between which no information is allowed to filter," explains the psychological expert witness in court.

*To keep those things that he neither sees nor accepts out of his consciousness, he resorts to various defense mechanisms, including denial, compartmentalization, and rationalization—for example, saying "I had no choice." Another strategy is isolation: keeping things at a distance and describing everything, whether you're talking about facts or emotions, in cold, factual terms, rather like a surgeon.*

We're all capable of compartmentalization. Compartmentalization is what allowed Duch to be a good father while knowing that children were being murdered at the prison. Once you've classified the other self as the enemy, it is possible to compartmentalize your two lives. One can go home and carry on as normal. The psychological expert witness continues, "We mustn't forget that we're talking about a man who had a very high idea of what he was doing—that is to say, serving Communism—and who was always convinced that everything he did was for the common good of the time, which is to say for the good of the Angkar."

**TWO OF DUCH'S BROTHERS-IN-LAW** were executed—one of them at S-21 under Duch's authority. He was the deputy head of the secret police in Kampong Thom province, and he was yet another Khmer Rouge torturer destroyed by his own Party. At first, Duch allowed his brother-in-law to write his initial confession unfettered, but then his boss, Son Sen, reprimanded him. Without a trace of emotion, Duch says: "I stayed calm. Later, he made many other unacceptable mistakes. If I had let him live, I would have endangered myself and my entire family. So I had him arrested, shackled, interrogated, and tortured."

Judge Lavergne asks, "With your usual attention to detail, I'm sure that you read his confession with great interest. What did your brother-in-law confess to?"

"I can only tell you what I remember. He confessed that he had been a member of an enemy network since before 1970 and that he had been tasked with marrying my younger sister after my imprisonment. That's what I remember."

"Which network was he supposedly a member of? The KGB, the CIA, or another one? Do you think his confession was credible? Did you believe it?"

"Your Honor, it is difficult for me to provide answers on this subject. I believed no more than 30 percent of it. Or maybe even less. Maybe only 20 percent."

Since his brother-in-law wasn't the head of the family, Duch managed to have his sister and her children spared. He hadn't vouched for his former teachers, or for his old friends, or his brother-in-law. But he did for his little sister. He promised to reeducate her. Judge Lavergne asks:

> *In* Revolutionary Flag, *there was much discussion of Party discipline and, in particular, of the admission or dismissal of Party members. And the following was written in the magazine: "It's impossible to reason with feelings. One must reason only according to Party principles." Is that a sentence that you've heard or read before?*

"This principle was applied by the Communist Party of Kampuchea."

Judge Lavergne asks the psychologist:

> *Could one say of the defendant that, because of his family history, his culture, his education, and his experience with Communism, he ultimately shut out all personal feelings, and that he could experience only those emotions that matched the Communist ideal or that corresponded to what society expected of him?*

"Yes."

Duch expresses his regret and remorse, but he doesn't suffer from depression.

"With Duch, there's a complete absence of a sense of guilt in the Western or psychoanalytical sense of the word," says the psychologist. "One could say that he could not then and cannot now feel guilt, since that implies a sense of empathy and it supposes that he stops compartmentalizing his sense of self and becomes self-aware."

# CHAPTER 14

**F**ROM THE TRIAL'S OUTSET, THE PROSECUTOR DECLARES THAT
Duch's regret is insincere and his account of events incomplete. Duch knew everything that took place at S-21, gave lessons in torture techniques, and supervised interrogations himself, says the prosecution. It was not the work of his subordinates: Duch was behind each case.

Certainly he reported to and received orders from his superior in the Politburo. But the prosecutor also claims that Duch was solely in charge of the concentration complex's daily operation and was not averse to using his independent authority. Duch ran the regime's most important prison and enjoyed direct access to the regime's most senior leaders.

The defendant says that his job was only to facilitate things; that it was the Angkar that decided whom to arrest, and his subordinates who decided whom to torture. Duch was merely passing on orders, he states. The evidence, voices the prosecutor, points to another conclusion: that the defendant brought about, supervised, and participated in the crimes, and that he knowingly sent people to their deaths. Duch, says the prosecution, controlled the torture machine of S-21 in its entirety. "These aren't acts carried out by a man under duress, but rather the choices made by a devoted revolutionary. We are convinced that the leaders have revealed only some of their crimes."

The prosecutor acknowledges that the defendant has cooperated with the court. He notes that Duch has foregone his right to silence and has facilitated the corroboration and discovery of evidence. He grudgingly admits that this limited collaboration has helped uncover some truths. But the prosecutor dismisses Duch's contrition as insincere and his confession as a sham.

Several times the prosecutor stares long and hard at the defendant. Duch sits facing the prosecutor, giving him his full attention. But when the trial ends eight months later, Duch turns his back on him.

ONLY A HANDFUL OF photos of Duch taken at the time of the crime exist. One shows a smiling Duch entering a room where a number of men and women dressed in black are seated around a big dining table. It looks like a prison-staff break room, but in fact the photo was taken at Duch's office, on the corner of Streets 25 and 310. The meal was no ordinary one; Duch explains that it was his deputy Huy's wedding day. The prosecutor hopes to use the photo to demonstrate that, contrary to what Duch has claimed, the prison director had close personal contact with his staff, including with guards such as Him Huy, who testified at the trial. The problem is that the prosecutor, though he has been working on the case for three years, has made an embarrassing blunder: he has confused Him Huy, the young guard who became a unit leader, with Nun Huy, whose wedding day is being celebrated in the photo. Nun Huy was number three at the prison complex and the person in charge of S-24, the "reeducation" camp that also fell under Duch's authority. Nun Huy was killed at S-21 a month before it was shut down. The prosecutor has just wasted forty-five minutes of questioning time.

With barely disguised glee, Duch watches his foundering opponent's line of attack collapse. Then, with great composure, he tells the court that, had the dinner been a commonplace occurrence, there would not have been a photographer present and, he adds, "I would have had my own separate table."

Duch finds nothing more invigorating than adversity. When he has an opponent on his knees, he is merciless. With the prosecutor falling to pieces, Duch shows the contempt that—as he knows better than most—is the condemned man's final revenge.

"The special unit: what did that do at S-21? What job did they have?" asks the prosecutor, still shaky after his blunder.

"I would like to remain silent about this."

"Why?"

"Because I have nothing to add to what I have already reported."

Duch dodges the prosecutor's best efforts to pin him down. Frustrated, the prosecutor takes refuge in a dead end. He and Duch are not on the same page. The prosecutor triggers the trial's slide toward a final, fruitless showdown. "I submit that you were not scared because you were extremely good at your job and you were a great asset for the Party."

"Yes, it is true that I was doing a good job for them."

"Your superiors were very satisfied with your work. You were a very proactive manager in implementing the Communist Party policy," charges the prosecutor.

"I told you truthfully what they ordered me to do. I had to follow the orders 100 percent."

"You were very proud of your work, of your techniques, and that you held that position."

"What I hoped was to stay alive, because I was so honest with them. They needed me and I was loyal to them."

"And that's why when you were in Pol Pot's company, it made you feel good. Do you remember saying that?"

"I never accompanied Pol Pot. I was happy because my former teacher, Son Sen, was the seventh member in the Party."

"You were one of the most highly connected Party members. You felt protected, you felt untouchable. That's why you were not scared. You inflicted terror on innocent Cambodian people."

"Who created that kind of paranoia? It was the Central Committee who imposed the terror."

After some rough days of testimony the previous week, Duch regains his footing with astonishing renewed vigor. Only a few days earlier, he had been at his lowest. His opponent's bungling inconsistencies bring him back to his feet. When Duch's fall finally does come, it will be of his own making. For the time being, however, he watches the other side with a smirk of amusement and contempt.

Duch has a biting and sometimes even entertaining sense of irony. For example, at one point, he illustrates how dispensing justice is really always a display of power, whether under the yoke of the Khmer Rouge or the banner of the United Nations. One of his crimes, he says, was to have inculcated his own notions, ideas, and methods into the young people he recruited and who had no choice but to accept them. He then adds under his breath: "It seems to me that the same thing is happening here."

Occasionally, he yields to the arrogance that feeds his sense of superiority. But he knows how to color it with humor.

"Mr. Kaing Guek Eav, could you please tell the court when you felt you were no longer the prisoner of the Khmer Rouge? In what year was it?" asks the prosecutor.

"It was on May 10, 1999, when the government put me in the hands of the military tribunal."

"Are you telling us that for the twenty-six or twenty-seven years since 1971, you had no chance to escape the Khmer Rouge? Is that what you're trying to tell us?"

"Could you please make sure your math is correct? Yes, I was detained by the Khmer Rouge in various ways."

Duch had been on the verge of falling apart; now he's sharper than ever. Faced with questions asked a hundred times, he defiantly asserts his right to remain silent. His answers can be so abrupt he sometimes sounds rude.

"I was in politics. Anatomy wasn't an interest of mine," he says in response to a question, posed for the umpteenth time, about the blood transfused from prisoners.

Only three months into the trial, the international chief prosecutor deems it appropriate to quit. At the ensuing press conference, he says: "In modern war-crimes courts, all the information is kept on file precisely to avoid disruption by changes in personnel. The only thing that matters is that there is someone who is in charge, legally, of making decisions, and who does just that. Everything else stays on course."

To Duch, that must sound like a description of S-21.

When the international deputy prosecutor sits back down in the courtroom, there's a terrible weight of powerlessness hanging over him. On June 22, he sinks into an abyss from which he does not emerge until six months later, during the final arguments—and then only because the defendant brings about his own downfall.

**ANY FOREIGNER WHO HAS ATTENDED** any sort of public forum in Cambodia—whether it is a public information session, a press conference, or a courtroom cross-examination—has walked away feeling discombobulated, exasperated, or even laughing out loud at the strange workings of the Cambodian mind in action, and its unique tendency toward repetition. During my first press conference there, the Cambodian journalist who took the microphone from me asked the exact same question I had just asked, word for word. Utterly taken aback, all I could do was stare at the journalist.

Just as they have a cyclical rather than linear notion of time, thinking progresses circuitously rather than in straight lines.

"They speak in circles while you speak in pyramids," filmmaker Rithy Panh told me.

Circular contrasts with linear, reiteration with accumulation, detail with summary, and the present moment with the past's calibrated and inexorable pull toward the future.

When a Cambodian judge makes a point, the reply he receives is accepted unconditionally, with no follow-up question. The journey from a specific question to an appropriate answer is often a long and winding one. The mind follows its own path, of course, but the Cambodian brain follows twists and turns that sometimes leave the Western brain, accustomed as it is to moving quickly toward goals, deeply distressed.

I was surprised to find that sometimes all the repetition eventually begat a totally new answer. It brought to mind Native Americans, who take the time to pass a ceremonial pipe around the entire group before anyone says something of importance. When I ask the psychologist who had examined Duch about it, she replies matter-of-factly that circular thinking was the way they worked in therapy, and that it is this attention to detail that allows the therapist and patient to understand the bigger picture. Therefore, she says, she's not bothered at all by the meandering way things are discussed in Cambodia.

The problem is that in a Western-style trial, repetition is not considered a way of reasoning but a professional failing, and throughout the trial, almost everyone involved complains about the exhausting redundancy of the arguments—the responsibility for which, by the way, does not lie solely with the Khmer lawyers. The president of the trial chamber, Nil Nonn, asks—repeatedly—that people cease repeating themselves. But though it is his job to keep order in his courtroom, there's little chance that the judge will solve the problem so long as he himself sees narrative repetition as perfectly normal. The moment when, on the forty-ninth day of the trial, one of the judges again asks a witness about some details that have been established a hundred times over—How were the prisoners chained? How did they wash? How did they relieve themselves? What did they eat?—is the point at which, for many of us, the culture clash reaches its nadir.

By early June, other things only add to the fatigue. At the start of the trial, hearing the defendant in his own words was one of its most interesting aspects; now it has grown tedious. The conduct of the proceedings has fallen apart. The presiding judge is so apathetic it's as though no one's in charge. The lawyers strive to outdo one another in the irrelevancy and repetition of their arguments. Duch is in control of the hearing and of his own case. He is his own prosecutor and his own defense and sometimes even his own judge. In fact, he so dominates the courtroom that it actually works against both him and his brilliant lawyer, François Roux. The sheer deftness of the master of confessions, combined with Roux's artfulness, leads to an odd paradox: instead of being outraged by the prosecution's ineptitude or the presiding judge's incapacity, people resent Duch and Roux for being well-prepared and efficient.

Then, on the day the prosecutor faints, the presiding judge at last measures up to his position. In fact, on June 22, Nil Nonn goes through a complete metamorphosis. From a weak, nonexistent, nervous, and mute president of the tribunal petrified by his own responsibility, he is suddenly transformed into an effective judge, one who even tends to be a little rash, perhaps, but who, most of the time, is reasonable and never excessive. It's as though the long and patient effort of repetition has finally borne its fruit of wisdom. Yesterday, the judge balked at settling even a basic point of procedure; today, first he gives a clear and pertinent answer to a dissenting lawyer, then pulls the civil parties into line, then brushes aside a prosecutor, who retreats behind his desk, and finally frees himself from the defense's domination—and all this without consulting his peers. Like anyone enjoying unrestrained power, Nil Nonn sometimes comes across as arrogant. But he is now firmly at the helm of a ship that had been in dire need of a captain.

When Cambodians of the old generation say "Khmer Rouge," they don't pronounce the final consonant, the *je* sound. Thus, they call Pol Pot's men "Khmer roo," which in French can be written "Khmer Roux." But when Nil Nonn addresses François Roux, he regularly calls him "François Rouge." One day in the middle of July, a tired, stressed, and irritated Mr. Roux makes a number of uncharacteristic mistakes. The presiding judge doesn't bother with either the cordial "Monsieur Roux" or the more comical "Monsieur Rouge." With a big smile spread across his square face, framed by thick-rimmed glasses, the judge upbraids the lawyer for the first time in the course of the trial. In a cold and firm voice, he says, "Counsel for the defense, do you have anything more to say? I do not wish to be interrupted."

# CHAPTER 15

**N**OT COUNTING THE VIETNAMESE, SEVENTY-EIGHT FOREIGNERS died at S-21, including one "Arab," five Indians, twenty-nine Thais, a Javanese, a Laotian, three Americans, three Frenchmen, two Australians, a Briton, and a New Zealander. But no list is complete.

The law recognizes the overwhelming extent of the Khmer Rouge's crime in the name it reserves for such atrocities: a crime against *humanity*. But it's the non-Khmer victims who help embody the universality of it. The court pays particular attention to the Western victims. Only a handful of parents of foreign victims—all of them European or American—applied to be included as civil parties in the trial. The Arab and the Javanese remain utter mysteries. And in the end, the only foreigner whose story is told in any detail is the New Zealander Kerry Hamill. Like doubtless many others, it's a harrowing and unbearable tale of bad luck.

Kerry Hamill was the oldest of five siblings, four of whom were boys. In an old black-and-white photo, Kerry is on the bow of a sailing dinghy, its single square sail raised, with two other boys around his age. Sailing is their passion. In another photo, taken several years later, Kerry stands shirtless, sporting a sailor's beard, his hair wet with seawater. The handsome, smiling youth in the photo (he was twenty-six) looks strong and healthy. It is 1978; Kerry and some friends have bought a yacht and are sailing around the world. He regularly writes to his family. For his youngest brother, Kerry's letters are extraordinary tales of adventure. In July of 1978, Kerry sends a letter from Singapore. His girlfriend, Gail, disembarked there to spend a couple of months with her family back home. In August, Kerry and his two friends cross the Gulf of Thailand. Caught in bad weather, they seek shelter near an island. Then, out of nowhere, a boat starts shooting at them. Stuart, a Canadian, is killed. Kerry and John, an Englishman, are hauled away. The boat was a Khmer Rouge patrol.

Kerry and John were locked up in S-21 and accused of being CIA spies. There's little doubt that they were tortured into making their confessions and young Kerry gave up to his torturers the names of all the traitors in his network. Most remarkably, he managed to make a mockery of their macabre parody without their knowing. His handling officers, colonels, captains, and majors were all named after his friends in New Zealand. Names Kerry snitched to the Khmer Rouge included Colonel Sanders, Sergeant Pepper, and Major Ruse. His CIA instructor was called "S. Starr," a nod to his mother, Esther. Despite being deep inside the dark and terrifying jail of the black-clad Khmer Rouge, despite the torture, the handsome sailor kept his sense of the absurd intact. After two months of incarceration, Kerry signed his "confession." He was wiped out, along with John, shortly thereafter. His family's pain was just beginning.

Christmas of 1978 came and went with no news of Kerry since Singapore. Sixteen more months passed before his youngest brother, then sixteen, learned that Kerry had been captured, tortured, and murdered by Pol Pot's regime, which by that time had been overthrown. Family arguments at home multiplied and got increasingly heated. Kerry's second sibling, with whom he'd been very close, sank into a deep depression. Eight months after he learned of Kerry's death, Kerry's brother threw himself off a cliff. He was the same age that Kerry had been when he died.

The youngest Hamill is forty-five years old when he takes the stand at Duch's trial.

"Duch, when you killed Kerry you killed my brother John as well," he says, looking directly at the defendant.

Duch gives a little nod. He is sitting straight, his forearms on the table, his eyes fixed on the witness. It's his usual posture of respect and attentiveness.

After losing her second son in a little over two years, Kerry's mother rarely left her bed. Her room was like a mausoleum, says Kerry's youngest brother. Depressed and suffering from shingles, she lost all interest in life. Kerry's father, meanwhile, retreated into himself. He stopped going to work and soon retired. They couldn't go on as parents.

"Our immediate family became a little bubble, and we became very reluctant to interact with others," says Kerry's brother.

For a while, the youngest brother lost himself to drinking. His schoolwork deteriorated. He couldn't stop the images from eating away at him. He imagined Kerry seated among piles of tires, being burned alive. That's how the Khmer Rouge erased every trace of the foreigners they killed—by burning them on street corners, though they probably killed them first. Then there's the haunting photo from the archives of S-21: a man on the ground with his feet shackled, lying in his own blood yet propping himself up on one arm, bravely trying to raise himself as the photographer releases the shutter. The photo isn't clear enough to give us the man's identity. But for Kerry's little brother, in tears, there's little doubt: "For me, this is my gorgeous, beautiful brother Kerry Hamill at S-21," he says. "This is the sort of image that haunted me when I was sixteen and still haunts me today. I have lost so much sleep over this image."

He stares at the former director of S-21. Duch sits up straight, looks at him and listens carefully.

"Duch, at times I've wanted to smash you, to use your own words, in the same way that you smashed so many others. At times, I've imagined you shackled, starved, whipped, and beaten viciously . . . *viciously*," repeats the little brother with venomous rage.

*I have imagined your scrotum electrified, your being forced to eat your own feces, being drowned, having your throat cut. I have*

*wanted that to be your life, your reality. I have wanted you to suffer*
*the way you made Kerry and so many others suffer. However, while*
*part of me has a desire to feel that way, I am trying to let go and this*
*trial is part of that. Today in this courtroom, I am giving you all the*
*crushing weight of that emotion—the anger, the grief, and the sorrow.*
*I'm placing this emotional burden on your shoulders. It is you who*
*should bear the burden alone. From this day forward, I feel nothing*
*toward you. To me, what you did removed you from the ranks of*
*humanity.*

The little brother has a few more questions to ask the defendant.
"Duch!"
Duch immediately gets to his feet.
"I acknowledge that you plead guilty . . ."
Presiding Judge Nil Nonn steps in. He tells the defendant to sit
down. The judge warns Kerry's brother that he doesn't want to hear
any abusive language. He doesn't want to hear any vengeful talk. He
tells the youngest Hamill that he must put his questions to the judge,
not directly to the defendant. Nil Nonn knows that some of his colleagues, unused to seeing victims of a crime participate in the perpetrator's trial in this way, are horrified that Kerry's little brother is
allowed to say the things he does in court. "I am angry beyond words
with you and what you did," continues the younger brother after Nil
Nonn's warning, "but I acknowledge and respect your guilty plea.
Your acknowledgment is a small but significant contribution to addressing the harm that you caused. You have proven to this court that
you have a very good memory. Please answer this question: what do
you remember of my brother?"
"There were four Westerners," says Duch in a slow, gentle voice,

*but I remember only a young Briton, John. He was very*
*gentle. I did not meet Mr. Hamill. He wrote in detail his complete*
*confession and I believed his confession. John and Kerry were*

*executed simultaneously and their bodies had to be burned to*
*ashes as ordered. The actual dates of the executions I cannot tell.*
*They would be taken away and smashed after the confessions were*
*extracted from both of them.*

Duch told investigators that he had entrusted the interrogation of Kerry and John to Pon, his favorite interrogator, who had a "good command of violence." One night, Pon came to tell Duch how nice and polite the Englishman was. The next morning, Duch went to see for himself. Pon wanted Duch to initiate the interrogation session, but the boss wanted to watch his subordinate in action. He was also concerned about the skill of the interpreter.

"What's your name?" Duch asked the person who had been recruited to interpret.

"Sarun Chon."

"How do you say *kaun mi sampheung* in English?"

"Son of a bitch."

"I knew then that he had all the skills," Duch told investigators.

It's better to avoid imagining how the rest of the interrogation went. No doubt it was more direct than circular.

Kerry Hamill's youngest brother says he isn't seeking any sort of financial restitution, which is just as well, since the court has no authority nor will to make any. But he asks Duch how he thinks he might atone for the harm he did the Hamill family.

"The best I can do," says Duch,

*is to get on my knees and pray for forgiveness. The victims and survivors can point a finger at me. I am not offended. It is your right and I respectfully accept it. Even if the people stone me to death, I won't say anything; I won't say that I am disappointed or that I want to commit suicide. I am responsible for my acts. It's for others to choose whether they forgive me or not. I am here to accept my responsibility. I am filled with remorse for what I've done. I mean that from the bottom of my heart. I'm not saying this as an excuse. I mean what I say.*

At that instant, I almost feel a connection between the two men. But this turns out to be just another one of those dreams that the trial abruptly shatters.

# CHAPTER 16

**I**T WAS SUOR THI'S JOB TO CHECK OFF THE NAME OF EACH PERson leaving through the prison gate. The prison clerk knew that if he made a mistake, he would be considered an enemy, and he was in a better position than most to know what became of the regime's enemies. The gate opened just enough that no more than one prisoner at a time could pass through. Suor Thi ticked off each name on the list. The prisoners' hands were bound, their eyes blindfolded. One by one, they were led aboard a truck parked in front of the gate. They used a chair that someone had put next to the back of the truck as a step. The truck was large enough to hold sixty people and there was a Land Rover available if more room was needed. The convoy left around six in the evening. The trip to the killing fields at Choeung Ek took about half an hour.

No one kept a list of the children sent to their deaths. The regime liked to recruit the young to do its drudgework, because they were "blank pages": when they were sacrificed, their "pages" stayed blank. According to Him Huy, the guard in charge of transporting prisoners from S-21 to the killing fields, the children were killed closer to the prison, in the center of town. But this wasn't his responsibility, he clarifies. Maybe so. Or maybe he finds it too difficult to confess to their executions. Peng and Phal, the other executioners held responsible for the children's murder, are both dead. The "others" are always dead.

Important prisoners—that is, high-ranking cadres purged from the regime—also received special treatment. They were executed near S-21 and buried near the intersection of Street 163 and Mao Tse-tung

Boulevard, says Duch, short of breath, with his mouth hanging open.

These special prisoners were hit in the back of the head and then had their throats cut, like everyone else. But, unlike everyone else, they were sometimes disemboweled and photographed. This was to reassure the Standing Committee that their former colleagues were dead. The only other victims who were photographed postmortem were those who died in the prison, prematurely, from torture. Prison staff photographed their bodies to prove to the director that the prisoners hadn't escaped.

Once they arrived at Choeung Ek, the prisoners were led to a wooden shack, one at a time. The generator was turned on so that there was light and, some say, to drown out the sounds. The executioners gathered around pits similar to shell holes, dug into the field around the house. They carried torches and the tools they needed. Him Huy would then go through the list of that night's victims, checking each name with each prisoner. He had to make sure that he had correctly checked off the names on the list he took back to the prison. An executioner would lead one prisoner at a time from the shack to an execution pit. The executioners would tell the victim that they were taking him or her to another house. The executioners tried to put the victims at ease, to make sure they died in silence.

"We told them to kneel by the pit. We hit them on the back of the neck with an iron bar. We cut their throats. Then we took off their handcuffs and clothes," says Him Huy.

The men in black killed by night. The executions began around nine p.m. and could last until dawn, depending on the number of people they had to kill. At seven the next morning, Suor Thi had to provide his superiors with the "list of destruction" containing the names and jobs of those who had been executed, as well as the date on which they had been destroyed by the Revolution.

**"I DIDN'T PAY MUCH ATTENTION** to the smashing. It was a practical issue," says Duch.

Yesterday, Duch broke down while listening to testimonies describing how things worked inside S-21. Today, the descriptions of the executions—quick, simple, almost mechanical acts that took place well away from the prison—help him recover. He goes back to speaking in that measured way that gives him self-control.

> *I saw myself as a senior police officer, not an inspector. Have you ever been in the army? When an officer needs something done, he sends a subordinate. He doesn't do it himself. You don't need to teach a crocodile how to swim: it already knows. There was no reason for me to go and inspect their work. I never thought about the method or the practical aspects of execution. Their job was to make sure the prisoners were smashed by whatever method necessary.*

Duch can describe how to make sure a victim is dead by cutting his throat. When he does, he lowers his voice. But just as he insists that he followed the interrogations and torture only from a distance, Duch also wants us to believe that he was in no way responsible for the executions. It was Duch's decision to create the killing fields at Choeung Ek, some fifteen kilometers outside Phnom Penh, because, he says, he was worried about an epidemic breaking out in town. It was one of his responsibilities as the head of the prison. But he saw no need and felt no desire to attend the executions. Just as his boss, Son Sen, only visited S-21 once, "on principle," Duch visited Choeung Ek twice, under orders. He claims to have only once seen executions with his own eyes, at five in the morning after a long, long night of killing at Choeung Ek.

Even within the death mill that was S-21, the actual task of killing prisoners was considered a lowly one. None of Duch's friends from the *maquis* nor any of his other protégés (such as the teenagers he recruited from Kampong Cham) were assigned to the execution detail. In court, Duch has a few compassionate words for those who, like Him Huy, had to carry out the loathsome chore. "I don't believe they acted without conscience or remorse. I believe they had such feelings.

I'm conscious of this; I understand it. I think everyone felt ashamed and remorseful about it."

HIM HUY'S "BIOGRAPHY," found in the S-21 archives, reveals what a twenty-two-year-old revolutionary's self-criticism looked like at the time. First, he wrote:

I did my best to carry out every task—no matter how humble or important—entrusted to me by the Party, and to carry it out without hesitation or objection, regardless of how difficult, demanding, or complex it was.

He then admitted that he still had a way to go:

I speak rudely to my comrades. I sometimes play the fool. I'm easily offended and quick to anger. As a leader of the masses, I'm not passionate enough. My observation and analysis of the masses are insufficient. I have not followed the activities of the enemy with sufficient rigor. I still underestimate the extent of our enemies' activities. I continue to be lazy when things need to be done immediately. I tend not to learn from my actions. I'm still too lenient. I still lack the concentration to lead the masses. Having taken stock of my faults, I want to express my determination to improve those parts of my personality that are still insufficiently revolutionary; I will constantly cleanse myself and build an unshakeable foundation from which to help build the Party.

Such was the ideology of the time. Like his boss, Him Huy has since disavowed it. "My biography, like many others, isn't truthful. We had to put what was expected of us in our biographies. I just followed the same template as everyone else."

Him Huy says that Duch's deputy, Hor, taught execution techniques. He also says that he twice saw Duch at Choeung Ek, in 1977. It's the most incriminating claim he makes against his former boss. Unfortunately, he gives several conflicting versions of what he saw. The testimony he gave during the trial's investigation phase was damning. He told investigators, "Duch accompanied people. There was one prisoner left, and Duch asked me: 'Are you determined or not?' I told him I was. He ordered me to kill him." But his testimony in court is much weaker: "Yes, I remember my statement. It was clearly my leader, Duch, but I had to rush to finish my job. At that time, I could not clearly say who was who."

"Did the accused order you to execute someone on several occasions?"

"I am not really sure now whether at that time it was Duch or Hor, because it was almost dawn and we were in a rush to finish the job."

The judge asks the defendant to stand.

"Was he the person who asked you to execute a prisoner?"

"As I just stated, we were in a rush. So I was not sure if it was Duch or not at the time. It was either him or Hor, because he was also present at the time, along with Hor."

Duch knows Him Huy well, and shows him a degree of respect that he denies Prak Khan, for instance. Nevertheless, he counterattacks by deploring the "shortcomings" of Him Huy's deposition. He keeps his eyes fixed on Him Huy, who leans forward like a penitent. The former prison boss lists all the elements of the former guard's testimony that he deems correct, and contests all the elements that incriminate him directly. Him Huy should not have said what he said about Hor and himself, says Duch, because "he does not know."

Him Huy cannot bring himself to look in Duch's direction. He limits himself to repeating: "When about a hundred prisoners were killed in one single night, Hor and Duch were there. They left before us. We were in such a hurry and worried that the work wouldn't be finished before dawn. I stand by my testimony."

Roux has no intention of letting such a threat hang over his client's head. He reads the following statement from the minutes of the on-site reconstitution of the crime: "The witness Him Huy has made conflicting statements about the number of times Duch visited [Choeung Ek]." The investigating judges noted that he said that Duch visited "from time to time," then "once or twice," then that he didn't know whether it was more than once. During the hearing, Him Huy says that he saw Duch twice. The first time, "the situation was chaotic; it was almost dawn, there was a man by the edge of the pit. The executioners were rushing to finish their job. That's why I'm no longer sure if it was him." The second time, "I didn't pay much attention, because he was still in his car." Yet another version of events, quite different from what he initially told the investigators.

"We'll leave it there for the chamber to assess," concludes the lawyer.

But the chamber chooses to assess nothing whatsoever. If the judges have any opinions about Duch's alleged visits to the killing fields, they keep them to themselves. Prudent in the extreme, they decide that there's enough doubt that they do not have to decide.

**ON THE SECOND AND THIRD OF JANUARY 1979,** the selection process for those to be transported to Choeung Ek was drastically expedited. The Khmer Rouge's leaders knew that the Vietnamese troops' rapid advance might force them out of the capital in the coming days. Brother Number Two, Nuon Chea, ordered Duch to kill all the prisoners. Duch says that he asked him to spare four ex–Khmer Rouge soldiers for interrogation. The four had been arrested following the recent murder of a Western journalist who had been on an official visit to Democratic Kampuchea. Brother Number Two granted Duch's request. Over the next two days, every other prisoner was executed. "I still couldn't believe that the Vietnamese were approaching. I thought that the prisoners were being killed in order to make room for new ones, like before. But I was wrong. Then I thought it was going to be my turn. I was exhausted. I couldn't work. I slept all the time."

Duch is an arduous worker, so when things go wrong for him he

claims he is "ruminating," which is to say he is overtaken by a paralyz-
ing self-doubt. He has twice experienced such episodes, he says: once
during the final weeks of S-21, and again two years later, when he was
living a precarious life in the *maquis* at the Thai border. According
to the psychologists before the court, such episodes are symptoms of
depression:

> *Duch's doubts and his uneasiness increased whenever the Angkar's*
> *line was no longer clear. A lack of clarity is extremely challenging for*
> *the psyche of someone with an obsessive personality. The result is a*
> *kind of slackening. Sleep becomes an escape, because you're looking*
> *for answers; when you're looking for a new way of living your life, you*
> *sleep. One can also speculate that the fall of the Khmer Rouge called*
> *into question all of Duch's work. This is why he slept so much: sleep is a*
> *symptom of depression.*

Like Duch, Him Huy says that he didn't like his job, but that he
had no choice. Contrary to Duch, he was quick to leave the Revolution
behind after the fall of the Khmer Rouge. In 1983, he was accused of
being the director of S-21 and thrown into jail, as was Suor Thi. A
few months later, Him Huy was sent to work in the rice paddies along
the Vietnamese border. He suffered no ill treatment and, ten months
later, was sent home.

At the end of his testimony at the international tribunal, Him
Huy is asked what he expects to come from it.

"It's like being born again—that's how we feel. We are among the
lucky ones who survived. We only want justice to be done."

Roux reacts immediately: "You have passed yourself off as a victim seeking justice. While I admit that the situation wasn't one of
your choosing, the fact is that the criminal machine only worked because leaders like you carried out criminal acts, or ordered them to be
carried out, at every step."

"I'm not sure I understand," says Him Huy.

"Along the chain of command, each person played a role, and

each person participated in a criminal system by obeying his superiors' orders."

"We had to obey orders or else we would be killed."

"Is that why you think of yourself as a victim?"

"Yes. We were all victims."

# CHAPTER 17

**T**HE REALITY OF THE TOTALITARIAN EXPERIENCE IS OFTEN GRAY. The woman on the witness stand today has come to honor the man the Khmer Rouge decided she should marry. A revolutionary soldier, he was killed at S-21 in 1977. She describes how she joined the Communist guerrillas in 1971 "because I was very angry about what we were suffering at the hands of the American capitalists and imperialists." She went into the *maquis* "to liberate the country from those people," and ended up with the rank of company commander in Democratic Kampuchea's victorious army. When the Angkar arranged her marriage, she and her husband were one of three couples married simultaneously. Conveniently for a woman who found it difficult to celebrate being married to a man not of her own choosing, the Angkar had a remedy: there would be no celebration. Festivities were considered bourgeois.

"It all happened very quickly," says the woman.

> That morning, we were told that the wedding would take place at two in the afternoon. I was shocked and asked why we were being married so quickly. I asked if my parents, my family, and the people from my village were invited. The answer was no. I wasn't happy about the way our marriage was celebrated, but the times were what they were. The time had been set, and I couldn't refuse. I was also told that we were in a special unit and that we weren't allowed to marry someone outside the unit; I was told that the Angkar was like our parents arranging our marriage, and that therefore we had to accept the arrangement made for us. I was very unhappy on my wedding day.

One year later, her husband was purged at S-21 and she was sent to S-24 for "reeducation." After the fall of the Khmer Rouge in 1979, she returned to her village, where her mother told her that it was because of her, the revolutionary, that her father was dead. She fell to her knees before an aunt and begged her forgiveness, but the aunt refused to give it. So today she says that she must reject Duch's apology in order to prove that she isn't Khmer Rouge, that she is loyal to the nation, and that she was "betrayed by that group," symbolized in her mind by Duch.

If we look beyond the anticipated punishment for the crimes committed at S-21, we see how they have torn apart Cambodian families; we see the terrible burden of family betrayals and insurmountable feelings of guilt.

Another woman takes the witness stand. She's wearing a burgundy-colored jacket over a white blouse and an elegant sarong typical of city folk. Her hair, touched lightly with gray, is cut short and neatly pushed back. A thin pair of spectacles rests on her nose. She is seventy years old, but looks younger. She pinches the hem of her blouse and nervously pulls it down. A victims' assistant puts a hand on her arm. Of the hundred or so students who passed the entrance exam for medical school in her generation, this witness was one of the few women. She immediately apologizes: "Sometimes I feel as if I am mentally unstable."

She speaks quickly and forcefully. There are notes in front of her, but she doesn't use them. As soon as she starts talking, her story carries her away. She describes how the entire population of Phnom Penh was evacuated in the hours after the arrival of the Khmer Rouge. She remembers each moment. She can still mimic the way black-clad soldiers with megaphones in hand insinuated that all educated people were to be eliminated. "They said that they would keep only the base people."

Her husband was deputy director of civil aviation at Phnom Penh airport. He was arrested. She was sent out to be "reeducated" by working on the dykes and dams. During the rainy season, her black-clad supervisor told her that if she passed this test, she would survive. If not, she would die. She closes her eyes to help jog her memory before diving back into the details of her tragic odyssey. Duch is sitting up straight, listening closely.

"I've lived in despair for so long that when death comes, I won't falter," she says.

After the fall of the regime, she returned to Phnom Penh and found work at the hospital. One day, her boss summoned her and told her to visit the museum at S-21. She knew very well the Ponhea Yat High School, where the Khmer Rouge had set up its detention center. Friends of her parents used to live close by. She reached the prison and was met by one of the survivors, she says. It's at this point that, in court, the pitch of her voice rises and cracks. Her speech becomes a series of short, strident cries, and she addresses the court in that striking timbre that the Khmer language reserves for anger, grief, and incomprehension. At S-21, she was shown documents, including a photograph. It was the last one taken of her husband, Thich Hour Tuk, alias Tuk. The documents contained the date he was brought to S-21: February 2, 1976, and the date he was executed: May 25, 1976.

In the photograph, the prisoner's piercing gaze appears to defy the photographer. He wears a thin mustache and has a few hairs on his chin. He looks slightly cross-eyed. Tuk is pursing his full lips, which gives him a skeptical expression. His brother, a pilot, was also destroyed at S-21.

The widow lowers her voice to give the court an impression of a conversation she had with a cousin, and another she had with a niece. Sometimes she seems disorientated and confused, as though suffering from the mental malady she mentioned at the beginning of her deposition. Then she reminds herself that the regime accused her husband of a "great crime." And then her angry voice returns and cracks through the courtroom like a whip and she asks the same question over and again: "Why? Why? Why?"

She says that men fall into one of two categories: those that resemble humans and have gentle hearts; and those that resemble humans and have animal hearts. An extremely devout woman, she prays for Duch's reincarnation and that "all of these beings cease to be cruel like Pol Pot's people." Then that question again: *Why?*

"Why should people who have done no wrong be locked up and mistreated? I don't understand."

Her story returns ceaselessly to the inexplicable, a circle without end: they came for him, he disappeared, he's dead. It is enough to drive you mad.

"This is a good moment to take a break," says the presiding judge.

**IT TURNS OUT THAT** it was the witness's older sister who denounced her husband to the black-uniformed guards. She considered her older sister like a mother.

> *We felt betrayed. She had been indoctrinated. That's why she said the things she did. Once, after all that happened, after all the suffering, I asked her what exactly Communism was. Now I know what it is: it's jealousy; it's competition and mass murder; it's sending people to S-21; it's betrayal; it's the denunciation of kith and kin; it's your loved ones getting arrested and executed. When I remember Buddhist teachings, I feel calmer; I understand that she did what she did because of the way the Communists brainwashed her. She denounced my husband. I blamed her, but perhaps she wanted to be Pol Pot's wife. She's the one who will have to suffer the consequences.*

The judges have fallen quiet. Her lawyer has cast her adrift on the river of her memory, aboard her raft of grief. Her lawyer doesn't ask a single question; not one person in the courtroom interrupts her frenzied torrent of words, her heartbreak, her pain and madness, and that question—*Why?*—that keeps coming back again and again, the woman banging her head against it until it bleeds. "I was loyal to my country. I was loyal to my husband. Why have I been punished like this?"

# CHAPTER 18

**P**OL POT, SAYS DUCH, WANTED TO BUILD A MONUMENT ON THE
Wat Phnom, the tiny hill in northern Phnom Penh where a
temple stands. This first sign of a personality cult—so typical of totalitarian regimes—occurred at the end of 1977, and ran counter to
the Khmer Rouge's obsession with secrecy. The regime needed painters and sculptors. Duch searched through his lists.

"Who in this cell knows how to paint?" shouted a teenage guard.

Bou Meng raised his hand. Moments later, he was in a room on
the ground floor of the prison. Someone handed him a snapshot
which, he saw, had been developed in China. The man in the picture
was unknown to him; nevertheless, Bou Meng was to draw his portrait. It was Pol Pot, the Revolution's "Brother Number One." Duch
sat cross-legged behind Bou Meng while he painted. If he failed, said
the prison director, he would be used for fertilizer.

"I didn't know if I was going to be used as fertilizer, or if I was
supposed to produce some," says Bou Meng. "It was a hard question
to answer in those days."

Duch gave him a sheet of paper on which to sketch. Satisfied with
the result, the prison director asked Bou Meng what materials he needed
to paint a large painting. He ordered his subordinates to fetch them.

Vann Nath had been lying in a cell in Building B for a month
when he, too, was called by a guard. Vann Nath was the last in his
row of prisoners bound in leg-irons, which meant the guards had to
unfetter all the others first before reaching him. He needed help to
stand; he was starving. He remembers thinking he was so hungry that

he would have eaten human flesh. He was led out of the cell, barely able to walk. He wasn't blindfolded. He was terrified, convinced he was about to die. He entered a building to find four people waiting for him, including "Brother East"—Duch. He was asked to summarize his experience as a painter since 1965. Bou Meng was already there. Vann Nath was told that the Angkar needed portraits. He replied that he would do his best. He was given a photograph of Brother Number One. Vann Nath had never seen him before. His ears hurt. He stank of shit. He wanted to shave his mustache. He promised not to commit suicide. He felt on the verge of fainting: if he didn't paint well, he would die; if he did paint well, he would also die, just a little later. He was told he could rest for three days. "I realized it was a matter of life or death. My first painting was a failure. It was in black-and-white, which was a new technique for me. I asked for colors."

A photo of Pol Pot appears on the courtroom screen and stays there longer than did the image of Bou Meng's wife. A murmur rises in the public gallery, followed by whispers.

At first, Vann Nath wasn't much good, but "Brother East" thought he could get something out of him. On February 16, 1978, Duch used his red pen to cross out Vann Nath's name from the list of people to eliminate. "Keep for use," he wrote in the margin.

One day, Duch brought in another prisoner and told him to make a sculpture of the mysterious Pol Pot. But it turned out the prisoner didn't know how to sculpt; he was just trying his luck. He was never seen again. Another prisoner, a Vietnamese man, claimed that he could make paraffin-wax molds. When he failed at his attempt, Duch got angry and had one of his interrogators hit the prisoner. He was taken away and never seen again. Next, a Japanese prisoner tried to save himself by joining the artists' studio, but it was also in vain.

**THE STUDIO WAS AT THE END** of Building E, the smallest, central prong of the former school's five buildings laid out like a trident. The entrance at one end of this modest house, which feels a little overwhelmed by the four large, three-story blocks surrounding it, was where Suor Thi met prisoners as they first arrived and took their photographs. The shutters on the windows at the other end of the building, where the studio was located, usually stayed shut. Vann Nath heard screams on an almost daily basis. He was shocked at first. But gradually, he grew accustomed to them.

Prison life improved substantially for the artists. Instead of miserable gruel, they ate rice, the same diet as the guards. "I even had noodle soup," says Bou Meng. They slept unshackled in the generator room behind Building E with four other prisoners taken from the cells "for use" by the prison's administration or the regime. They could hear the trucks coming and going from the gate, but they couldn't see anything. The artists were locked in their studio. There was no guard in there with them, but the system was set up so that each prisoner knew as little as possible, and nothing filtered out.

Bou Meng requested three months to produce a three-by-five-meter portrait of Pol Pot. Duch ordered him to fix the leader's throat, telling the artist to get rid of a lump that looked like a tumor. When Bou Meng mentions this in court, Duch gives a slight smile. He also instructed Bou Meng to fix the lips. "I survived because I was able to paint a faithful portrait of Pol Pot."

For Duch, who had taken his revolutionary name from a sculptor, the studio became a place of refuge. He visited almost every day. Each time he entered the room, the artists had to move to its far end and wait for his orders. Though he appreciated their work and often complimented it, the artists remained frightened of him. He never hid his displeasure when they fell behind schedule. Still, Vann Nath couldn't bring himself to believe that Duch was capable of putting people to death. "He was an intelligent, attentive man. He made you aware of his power. But he never did anything whatsoever to frighten us. He was respectful."

DURING A TRIAL RECESS, Duch holds a long conversation with his Cambodian lawyer, Kar Savuth, who, as is his habit, is resting on one of the chairs in the courtroom. Duch gives a short laugh, revealing his famously bad teeth. He lifts his head, and for a moment his bright, wide-eyed gaze freezes. Then, apparently in a talkative mood, he turns away to talk to his French defense team.

While Vann Nath testifies, Duch's expression is inscrutable, as though he has withdrawn into his shell. He still has some influence over his former subordinates, but he has no authority whatsoever over the painter. Vann Nath rubs his belly and kneads his handkerchief. With his jowls and heavy eyelids, he looks like a sad tortoise. Throughout 1978, the painter tried to produce his best work and to do what was asked of him, he says in his deep and slightly nasal voice. He had only one goal: "To survive." One day toward noon, Vann Nath got the feeling that they were being watched, and had been for some time. He felt a surge of fear when Bou Meng was summoned. Bou Meng left the studio and didn't come back. Vann Nath says he naively thought that his fellow artist had been released and sent back to a co-op. But two weeks later, he heard someone call out and the sound of chains scraping along the ground. In the doorway stood Bou Meng in chains, his skin pale and his hair long. Brother East stood behind him. "He said, 'a-Meng, what did you promise? Get on your knees and apologize to all of us.'"

With typical gallows humor, Duch asked whether Meng was still of any use or whether he should be turned into fertilizer.

Duch pays close attention to Vann Nath during his testimony. He listens carefully with his mouth slightly open, as though astonished by what he is hearing. There aren't many witnesses like Vann Nath: sober, firm, clear, and scrupulous, always taking pains to point out when he personally witnessed something and when he didn't. His testimony is both factual and charged with emotion. He exudes a natural dignity. For the past three decades, Vann Nath has devoted much of his time to advocating on behalf of S-21's fourteen thousand victims. He has acquired a unique stature in the country. Even when Duch disagrees with what Vann Nath remembers, he doesn't challenge the painter.

But Bou Meng doesn't recall the incident, or if he does, he remembers it differently. "My memory isn't in perfect shape," he had warned at the start of his testimony. Of the three still-living survivors of S-21, Bou Meng, with his hearing aid and virtually toothless smile, is the one who bears the most obvious scars of the violence he suffered in the prison. Several times, he is reminded in court of that serious incident in the studio. But he fails to understand what he's being asked, or else he dodges the question, or says he has forgotten, or even flat-out denies it ever happened.

Bou Meng and Vann Nath are two honest men who share a deep bond of solidarity after having endured hell together, who saw and heard the most terrifying sights and sounds of their lives together, yet whose memories disagree on whether one of them was abused on that day, or not, or to what degree. Bou Meng sums it up enigmatically: "Even an elephant with four feet sometimes collapses."

We don't always need to understand everything.


**BOU MENG, THE CONTEMPTIBLE *A*-MENG,** with his disproportioned, broken, stunted, almost-deformed body, his back and shoulders covered in scars, his toothless face, his pierced eardrums, says what he knows, says what he doesn't know, and says what he thinks he knows. Those years of terrifying lies and slander seem to have inured him against taking the truth lightly. They made him a man of principle, one who still bears the scars of a savage and bloody scheme.

"I wouldn't seem so old if I hadn't been tortured," he says.

Then, having taken stock of his own state, he pokes fun at his friend and fellow survivor, who is ten years older: "But Chum Mey is still young!"

Bou Meng has a disconcerting ability to laugh at the slights—even the well-intentioned ones—of his fellow man. Like when he describes how he was electrocuted, and a lawyer representing the civil parties asks him, "How long were you unconscious?"

Bou Meng bursts out laughing. "You can't tell how long you've been unconscious if you're unconscious!"

This humble man is ridiculing the smugly superior lawyer; a cheerful sound rises from the public gallery.

Bou Meng doesn't recall the episode in the studio, but he remembers other abuses. He hasn't forgotten the day when Duch, while visiting the imprisoned artists, ordered him to fight an ethnically Chinese sculptor with a plastic pipe.

Bou Meng remembers this brawl clearly: a year before the start of Duch's trial, he tried to recreate it for the investigating judges, when they were reconstructing events at S-21. A crowd of people had gathered in the former artists' studio; Bou Meng grabbed a white plastic chair, put it in the center of the crowd, and mimicked the way Duch sat when he used to watch Bou Meng paint: leaning back comfortably, one leg crossed squarely over the other. Bou Meng smiled at his own performance and pointed a finger at Duch to show how the jailer used to offer him cigarettes. This unorthodox but powerful gesture epitomizes the influence the former prisoner now holds over his jailer. A short, twitching laugh lights up the crevices of Bou Meng's face.

"And then what happened?" asked the investigating judge.

"He ordered us to hit each other in turn. I don't know why."

"And you did it?"

"Yes, yes," said Bou Meng, smiling.

"Did you hit hard?"

"I don't have any scars left from it, but it hurt a lot."

"Is that true?" said the magistrate, turning to Duch.

"Yes. I told them to hit one another."

"Why?"

"I've forgotten the reason."

"Was there no particular reason?"

"I don't remember. Perhaps there was no particular reason."

When the psychologists interviewed Duch about this incident, he said he wanted to meet with his confessor first. If there ever was a reason, it remains a mystery.

Duch wasn't the only one who enjoyed humiliating Bou Meng. The guard Him Huy also wasn't able to resist indulging in a little prison-camp humor in front of his fellow guards.

"Let me put it this way," he tells the court, embarrassed. "Everyone would make comments about his size, about how small he was; they used to mock him and ask him how he managed to have a wife."

One day, the young prison guard's mocking went much further. Him Huy heaved himself onto a-Meng's shoulders and rode the prisoner like a horse, goading him to prove his strength.

"You found this funny?" says Roux.

"Back then, we used to chat among ourselves. I got on his shoulders to see if he could carry me. I just wanted to test his strength, that's all."

"Do you think he found it funny?"

"He used to say that he was strong and that he could carry me. I didn't threaten him in any way."

Bou Meng rubs his head.

"I didn't mean to harm him. I did it just for laughs," adds Him Huy.

Bou Meng says that he's delighted to have had the opportunity to testify. He feels relieved, lighter. He believes that justice will be done for the millions of victims of the Khmer Rouge. The only question he wants to ask Kaing Guek Eav is: "Where was my wife killed? At S-21? At Choeung Ek? Or somewhere else?"

Bou Meng needs to know in order to go there and pray for her.

"I would like this affair to be resolved quickly," he adds.

Duch gets to his feet.

*Mister Meng, you have touched me deeply. I was shocked when I saw you again in February 2008. I wish I could answer your question, but the answer lies beyond my knowledge. My subordinates were in charge of these things. I can only presume that your wife was killed at Choeung Ek. Please accept my highest consideration and respect for your wife's soul.*

Duch's face tenses up into an expression set somewhere between astonishment and grief—his awkward way of showing distress. He turns his head bashfully and sniffs. He leans on the desk, his arms trembling. He sits down. Bou Meng is holding his head in his hands. He rubs more balm into his forehead. Bou Meng is a tiny man, his body crippled by torments inflicted by both nature and his fellow man; the pain of his memories and past suffering proves too much, and he, too, breaks down. This flood of emotion flusters the presiding judge, who tries to fill the silence with a steady flow of empty words. There's a new sadness in Duch's eyes, one that first appeared a few weeks earlier, when another victim was mentioned.

**FROM OCTOBER 1978 ON,** the prison was a lot quieter, says Vann Nath. The armed conflict with the Vietnamese meant the regime had to put a stop to its internal purges and concentrate on fighting the external enemy rather than the one supposedly lurking within. On January 3, 1979, the last truck returned from the killing fields, and the prison was empty.

"The interrogations stopped. We were idle. We stayed inside the compound. All the prisoners were liquidated," says Prak Khan.

All except the ones that had been "kept for use," such as the painters and Chum Mey. After his twelve days of torture, Chum Mey was led back to his cell on the top floor of the south building. The S-21 leadership realized it needed a handyman. Chum Mey, who felt as though he'd been in death's antechamber, now found himself working in the mechanic's workshop. His job was to repair the prison's machines, including the typewriters used to write the confessions his fellow inmates made after being electroshocked. Chum Mey shared his bunk with Thoeun, the dentist. There were four or five of them working in that other workshop, the one Duch never visited, behind the prison's main buildings.

Toward noon on January 7, one year to the day after he arrived at S-21, Vann Nath heard gunfire. He and a dozen other prisoners gathered in the studio. A handful of guards came to fetch them and ordered them out. They marched single-file, trembling with fear, across Tuol Tom Pong, known today as the "Russian Market." They passed Prey Sar, the S-24 reeducation camp linked to S-21; they marched through the night, were separated from one another, then found each other again in the morning, when they reached National Highway 4. At Prey Sar, Chum Mey, who had been arrested only two months previously, was miraculously reunited with his wife and their newborn baby. They fled together. A Vietnamese military convoy appeared. Gunfire erupted. The S-21 guards ran off. The group scattered. Vann Nath found himself on the roadside, terrified, along with three other ex-prisoners. They wanted to go back to Phnom Penh, but they feared that the Vietnamese would kill them, and they didn't return until January 10. Chum Mey, meanwhile, had gone his own way with his family and another S-21 prisoner who had been spared "for use." They were captured by a band of Khmer Rouge. Another gun battle erupted, and Chum Mey's wife and the fellow prisoner were shot. Chum Mey also lost his baby while fleeing. He went back to Phnom Penh alone and found Vann Nath, Bou Meng, and four other survivors of S-21.

They were free. Chum Mey's and Bou Meng's wives and children were dead. Vann Nath's wife survived. Their two children, one five years old and the other six months, did not. Thirty years later, these three men are the only living survivors who can bear witness to what took place at S-21.

"I tried to let go," says Vann Nath,

> but the suffering I endured there isn't easy to forget. My memories still haunt me. I don't think I'll ever be able to forget what happened to me. I never imagined that one day I would be sitting in this courtroom. It's a privilege and an honor. I want nothing more. Usually, plaintiffs ask for some sort of reparation. But I'm not asking for any compensation. I want only one thing, something intangible: justice for those who died. That is what I hope this tribunal will bring. Sometimes, I grow tired of telling people how I suffered. But then I remember that

*it's about revealing the truth, about transmitting it to the younger generation, and my weariness is lifted.*

The back of the painter's skull is unusually raised, which adds to the grace of his height. He slips on a navy jacket against the air-conditioned chill. Leaving the courtroom with dignity and confidence, he gives the impression of being a visitor from some mysterious, rarified world of sages.*

---

* Vann Nath died on September 5, 2011.

# CHAPTER 19

**D**UCH FIRST BECAME INTERESTED IN POLITICS AT THE AGE OF fifteen. He heard about the Revolution in China, where Mao Zedong had seized power eight years previously, in 1949. The Chinese premier, Zhou Enlai, visited Cambodia in 1956. Duch, who has Chinese ancestry, drew a certain pride from this visit by an eminent Chinese politician to a Cambodia in which the ethnic Chinese sometimes faced discrimination. One of his teachers, Ke Kim Huot, gave him a few books about the situation of peasants and workers. Cambodian independence, recognized by France in 1953, was still fresh in people's minds. King Norodom Sihanouk was a major figure in the inchoate Non-Aligned Movement. North Vietnam's Communists had just won a stunning victory over the French colonists. Anything seemed possible. Duch read everything handed to him and everything he could get his hands on. His reading wasn't confined to politics—he was interested in Buddhism as well. Duch needed to admire people in order to act and to move forward in life. He sought masters and mentors, and the first of these were Buddhist monks, whom he respected and with whom he lived and studied.

At this decisive age, teenage Duch became aware of his family's social standing and of how men economically exploited others: his father, for example, was indebted to a moneylender uncle. Kaing Guek Eav began to discover, through his reading and thanks to certain teachers, that there existed an alternative system, one that claimed to eradicate the exploitation of man by man. Communism, then ascendant in many places throughout the world, seduced a number of young Cambodian intellectuals who were in revolt against the social inequality, corruption, injustice, and authoritarianism of the monarchical system.

In 1958, Duch, along with two other boys and two girls, founded a little study group. Duch was very close to one of the girls, Sou Sath. In court, it takes Sou Sath all of two minutes to paint a verbal portrait of Kaing Guek Eav: he was a kind and generous young man; a good student; someone with no secrets, who gladly shared his knowledge; someone whom everyone in the classroom knew but who had only a few real friends; who supported others in their studies; who never fought with anyone; who respected his teachers; who never skipped class.

A faint, almost affectionate smile appears on Duch's face. For a few moments, he's distracted by two latecomers making their way through the public gallery. He turns his gaze back to his old friend. Sou Sath doesn't understand one of the questions she is asked. She smiles. Duch does, too. An ancient, calm breeze blows between them, a tender air from the times past, the scent of an old and innocent friendship. Sou Sath, a retired teacher and former activist with the Cambodian League for Human Rights, resurrects, with her confident and frank words, the man who existed before Duch, the man he was before this damned Revolution.

"He wasn't the only good student. I was, too. He wasn't the leader of the study group. There wasn't one," she says cheerfully and confidently.

Duch, a prudish man, confided that he worried that frivolous games would sidetrack the study group. The group members therefore decided to address one another as "brother" and "sister." Sou Sath was called "aunt." This was to avoid any "sentimental feelings that might lead to amorous incidents." Sou Sath says that she was unaware at the time that Duch had suffered a romantic disappointment. She laughs about it now and glances at him. He laughs in turn. Their shared world has remained intact, fresh, and bursting with life like a rice paddy after the monsoon. She asks if she can see him in his cell after her deposition. Ten minutes at most, she says.

Huot Chheang Kaing doesn't testify before the tribunal. They were classmates for three years, but he hasn't seen Kaing Guek Eav since 1961. He comes to his old school friend's trial for one day and one day only. During a recess, he goes and sits among a few journalists in the press room. He still speaks good French, but the interview takes place mostly in Khmer. He recalls a talented child, "always top of the class," always first in math, physics, and chemistry. Kaing is cheerful. He liked to tease his classmate about his name. He remembers another joke they shared, but it gets lost in translation—either that or it requires a certain sense of humor.

In any case, here's the joke as it was relayed to me: "Duch used to say that drinking water before eating was good for your health. I used to say the opposite: if you drink water, you won't eat much! So everyone takes care of his own business."

One of the most pleasurable things about listening to stories of people from other parts of the world is when you no longer understand anything. During the first few months I spent covering the Rwandan genocide trials, a number of Rwandan witnesses used a proverb that clearly held some powerful meaning: "When a snake is wrapped around the calabash, you have to break the calabash." I thoroughly enjoyed watching the judges' faces as they scratched their heads and tried to make sense of this ominous adage.

Kaing wins you over by lacing his wisdom with humor. For example, he tells a highbrow pun about Communism he and his friends used to make: in Khmer, the word *communism* sounds a little like *kum menuoh*; *menuoh* is the collective word for man or humanity and *kum* means resentment. Therefore Communism, they used to joke, means "resentment against man."

From the start, Kaing disliked Communist ideology. Not everyone believed the Revolution would deliver what it promised, he says, which doesn't mean they were some sort of vile reactionaries. He had noticed how the Chinese were starving to death while more democratic governments provided life's necessities. He also found that the sciences were more advanced in liberal countries. He used to argue with his classmate Kaing Guek Eav. The students were split into two tendencies—communism and liberalism, "progressives" and "imperialists." Kaing also recounts that some French teachers had no qualms preaching their politics.

Duch was neither talkative nor funny, says Kaing. He was a very serious boy. His classmates found him a bit effeminate, but no one teased him, because he was such a good student. "After 1979, I was told that he had been the director of S-21. I didn't believe it, because he had been so gentle. Then when I saw the documents, I believed it."

The memory makes Kaing cry.

"It's the fault of the Democratic Kampuchea regime," he says, regaining his composure. Kaing evaded the men in black by hiding his education while toiling in the co-ops. He still considers Duch a friend, though he makes sure to point out that he thinks Duch deserves to be tried. When Duch catches sight of his former classmate in the public gallery, he makes his way over to the thick, soundproof glass during the recess and waves at him. He smiles, clearly delighted to see his old friend.

"When Duch approached me, I saw the same man that I knew back then. He hasn't changed. It was the same face. What's changed is that he used to have quite a feminine character. Now he behaves more like a Frenchman. He's firm," says Huot Chheang Kaing with a twinkle in his eye.

**WHEN KAING GUEK EAV** was admitted to the prestigious Sisowath Lycée in Phnom Penh, he became aware of the gulf between the living conditions of rural Cambodians and those of wealthy city dwellers. In 1962, at the Pedagogical Institute, he met a professor who had been educated in France and who was already a secret member of the Communist Party, Son Sen. Henceforth, Duch addressed him as "master." At the same time, a French professor of geography was teaching Duch a few Marxist principles. They resonated strongly with the young, idealistic student eager for social change. Through another teacher from the former colonial homeland, Duch discovered Stoicism, which teaches indifference toward anything that affects emotions.

According to the psychologist expert witnesses, "Duch acquired a sense of social devaluation very early. He tried to compensate for it with study and hard work; he adopted highly idealized male role models and endlessly sought their approval. Their recognition made him feel like he had his own identity, which he based on theirs."

That same year, 1962, the Sisowath Lycée was gripped by an intense protest movement. Sihanouk's police reacted by brutally repressing it. Chhay Kim Huor, a teacher whom Duch admired, was among those arrested. Though Duch played no part in these events, they left him deeply shocked and fanned his revolutionary ardor. One of his teachers warned him that joining the Revolution was like being inside of a circle: once you're in, there's no way out. But by now his faith burned fervently, and he could not put off his decision to throw himself into the roiling waters of the Revolution much longer. He decided to join in 1964.

"A slave society becomes a feudal society, which becomes a capitalist society, which becomes a socialist society before finally becoming a communist society," recalls Duch. "We started to appreciate this theory while studying elementary mathematics. 'From each according to his ability, to each according to his work; from each according to his ability, to each according to his need.' I really liked that theory. I believed it. I wanted a society based on that slogan." A society based on absolute abundance and an end to the problem of production. In essence, a utopia.

Duch completed his studies. In 1965, he became a math teacher in Skoun, less than a hundred kilometers north of Phnom Penh. He paid a visit to Sou Sath. He wanted her to join him in Skoun and teach there, too. But she didn't follow him. Judge Lavergne and Roux share a complicit smile but keep mum about what's behind it.

Duch gave up his math books and embraced Marxist and revolutionary literature. The first such book was an illustrated Chinese work with captioned photos. Then he bought *Everything Is Done for the Party*, the story of a Chinese mine worker who dedicated himself to the Revolution first in a weapons-repair factory, then in a bayonet factory, then in a heavy-weapons factory. The Chinese man was wounded in his eyes and hands and sent to the Soviet Union for treatment. When he came back, he was made professor of industrial drawing at the university. "I thought that if that was the way of the Revolution, I had to live up to it; I had to be capable of following it."

He devoured Georges Politzer's *Elementary Principles of Philosophy*—"published by Éditions Sociales," specifies Duch—as well as Mao Zedong's *On New Democracy*. He was completely fascinated by class warfare. "Every kind of thinking is stamped with the brand of a class," he repeats to the court in French. Another idea has lingered in his memory: to truly love the people is to sacrifice oneself in order to bring about the total dictatorship of the proletariat. Duch found the alternatives lacking. For example, Jesus Christ taught that, should someone strike your right cheek, you should offer them your left. At the time, Duch found this to be at the very least inefficient, if not outright idiotic.

"I didn't know how you could serve the people with that theory," he says soberly.

He read Gandhi. But Gandhi seemed impossible to follow, because he was half-human, half-divine, he says. Marx, Lenin, and Mao felt much more familiar to him. He was particularly seduced by the Chinese leader. He bought Mao's book of thoughts on conflict, *Four Essays on Philosophy*. He remembers several of the book's chapters, including *Where Do Correct Ideas Come From*, perfectly. "At the end of his book, Mao wrote, 'Let a thousand flowers bloom and let a hundred schools of thought mutually complement one another.' I loved that sentence . . ."

Later, in 1976, Duch tried to study Stalin's book on Leninism, the famous book by which young Eastern Europeans who had fallen under the Soviet yoke were supposed to learn Russian. But Duch gave up on it. Maoism, he says, remained his major intellectual influence.

# CHAPTER 20

"TEACHER! HELLO, TEACHER! THAT'S MY TEACHER!" THE MAN calling out is a tall sixty-year-old who, when he arrived in the courtroom, greeted everybody with his hands pressed together and the broad, charming smile of someone who has spent his life working in the fields. Duch was his teacher between 1965 and 1968, and the man has clearly kept a happy memory of a down-to-earth, scrupulously fair, and kind teacher who gave free private lessons to poor students and who didn't preach politics in the classroom. The man remembers hearing Duch mention communism at the end of a class, but without pushing it on anyone. "He was a good model. The students liked him. We gladly attended his classes."

Another student of Kaing Guek Eav, who became a secondary-school principal before retiring, describes the same gentle and accessible teacher.

"The way he talked to us encouraged us to be good students and to help one another. We could consult with him at any time," he says, punctuating each sentence with an odd, quick exhalation through his nose.

Kaing Guek Eav talked with his students about morality, about loving the poor, about acquiring knowledge in order to better serve the nation. He encouraged them to work hard, and led by example. One former student after another, both in the courtroom and among the public, painted the same picture of a simple, fair, accessible man who was strict but never cruel.

Even then, Duch was secretly supporting the Revolution by giving most of his salary to the movement. Of his monthly salary of seven thousand riels, he says, he kept only a thousand for himself. He gave nothing to his parents. Everything had to be sacrificed to the Revolution. He quietly ran a clandestine network whose members included a certain In Lorn, alias Nath, who, ten years later, would become the first director of S-21.

Duch may have decided to join the Revolution in 1964, but it wasn't until 1967 that he fully dedicated himself to it. At the beginning of that year, a peasant revolt erupted in Samlaut in the northwest of the country. The government put it down mercilessly. The far-right element then in power initiated a "hunt for Reds." Midway through the year, Duch put himself through secret training, during which he met Vorn Vet, one of the top leaders of the Cambodian Communist movement. Vet would eventually become Duch's direct superior in the *maquis*, as well as Brother Number Five or Six of the Politburo, before he, too, met his end at S-21—at Duch's hands. Duch's other victims at S-21 included his former teachers Chhay Kim Huor and Ke Kim Huot, as well as Nath.

By the end of 1967, it was time for Duch to bid farewell to those close to him. He visited his family and told them, a few friends and the person in charge of the pagoda, that he was going into the *maquis*. He paid a final visit to Sou Sath and her husband. It was October 21, 1967, he tells the court.

Kaing Guek Eav went into the forest in the Cardamom Mountains, in southwest Cambodia. On November 25, 1967, he stood before Ke Pauk and took an oath of allegiance to the Revolution and the Party. Ke Pauk later oversaw the massive purges in the north of the country that kept the S-21 killing machine running at full steam.

"Did you accept that political violence was necessary when you joined the Communist Party of Kampuchea?" asks Judge Lavergne.

"No one told me at the time that political violence was the Party's daily bread. I only found out later, when I was forced to become the director of M-13."

Forty years later, Duch rediscovers his fervor when he recalls taking that oath in the heart of the jungle. To demonstrate to the court the revolutionary salute, he stands sentry-straight, bends his elbow at a right angle and holds his implacably clenched fist level with his head. It is a gesture he faithfully performed daily for years and years, and it comes back to him with ease. His tightly clenched fist, his ramrod posture, and the way he holds his arm straight against the side of his body all bear witness to the burning conviction that consumed so much of his life. "Raising the fist like this signified that you mustn't betray the cause. I didn't betray it. I walked the straight line."

Comrade Duch had barely started his revolutionary career when he suffered a serious setback. On January 5, 1968, twelve days before the Khmer Rouge started their armed struggle, Duch was arrested by the police force of the regime he hoped to overthrow. He was found guilty of breaching state security and of consorting with the enemy. The day of his hearing, he spoke without a lawyer. His trial lasted half a day. He was sentenced to twenty years of hard labor. Duch did not appeal; being a revolutionary demands the utmost sacrifice and radicalism. Duch was locked up in the central prison, where he got to know a number of militant members of the clandestine movement. In May 1968, he was transferred to Prey Sar prison. Eight years later, Prey Sar, by then known as S-24, came under his authority.

Duch tells the court that the authorities at Prey Sar used to terrorize the inmates. Some prisoners were summarily executed, he says, though he didn't witness this directly. Some were tortured. Duch dryly reminds the court that prisoners were tortured under French rule, under Sihanouk's rule, and under Lon Nol's rule. "Therefore my experience was a combination of all these, even if I taught myself."

Duch was insulted, but never tortured. Nonetheless, it was within the walls of Prey Sar that he came to believe that torture was "inevitable." When a judge asks him whether such practices strike him as normal, acceptable, or outrageous, Duch stalls, unable to find an answer.

"As a revolutionary, I was prepared to submit to torture. I wasn't frightened. I knew what was to come. I joined the revolution to change society, to transform it, to oppose the government, and to end government-sponsored torture," he says calmly.

There's a determined look in his eyes, though they've lost their usual strange gleam. Duch wants to convince the court that his spirit of sacrifice was undiminished. Then he tries to tackle the question: "If we had to judge it . . . we knew it was a crime . . . but how could we oppose it?"

The question is put to him again: "Were the torture and executions criminal acts, yes or no?"

"I was aware that they were crimes, and I knew that we had to fight for the Revolution. But I don't want to hide behind events. You are the judge of me."

PREY SAR UNDER SIHANOUK wasn't like Prey Sar under Pol Pot. For one, there was a doctor in residence. For another, the food was better. Reading was allowed. Duch says that he even continued studying Mao—clear evidence of how permissive Sihanouk's security service was. Family members were allowed to visit on Thursdays. Ultimately, you could get out alive.

On April 3, 1970, after two years and three months of incarceration, good fortune smiled on Duch. Two weeks previously, Lon Nol, Norodom Sihanouk's army chief of staff, staged a coup while the prince was abroad. One of Lon Nol's first actions was to announce the release of almost five hundred political prisoners. Duch had already been tried and sentenced, but a distant relative of his mother's had ties to Lon Nol. Duch was released.

The coup of March 18, 1970, determined Duch's fate and changed the course of Cambodian history. It amplified the terrifying bombing campaigns by the Americans. From 1969 to 1973, at least 540,000 tons of ordnance were dropped—blindly and from high altitude—on Cambodian territory. By contrast, a "mere" 160,000 tons of ordnance were dropped on Japan during World War II. The putsch gave the Khmer Rouge a huge, undreamed-of boost when Sihanouk backed the nascent guerrilla movement, essentially legitimizing it in the eyes of many Cambodians. It also tipped the country into an outright civil war in which around 600,000 people died between 1970 and 1975.

Duch explains with reasonable clarity:

*Sihanouk was the head of state. He used populist politics to defend the monarchy. Lon Nol was subservient to the United States. If Nixon hadn't recognized Lon Nol and if the Khmer Rouge hadn't cooperated with Sihanouk, the Khmer Rouge insurrection would never have succeeded. Sihanouk said that all Cambodians should go into the* maquis *and fight. That is how the Khmer Rouge got its support.*

Duch and his cellmates Mam Nai, Hor, and Pon regained their freedom. These men, who had devoted themselves to the education of others, and who, before they embraced Leninism, had been educated in the spirit of the Enlightenment, went on to become the operators of a merciless machine that ground other men into dust. Hor became the deputy director of S-21; Pon and Mam Nai, both teachers, became chief interrogators under the authority of Duch—also a teacher, and more talented than either of them. All followed orders handed down by their master, Son Sen—another teacher. No amount of education has ever inoculated a person from violent behavior, not even the most extreme variety.

After his release, Duch spent three weeks at home. Then he went to a monastery. Shortly after, he resumed his revolutionary activities. Four months later, in August of 1970, he obtained authorization to enter the "liberated zone," the part of the country already controlled by the Khmer Rouge. Now Duch's revolutionary life began in earnest.

**"HOW DOES A PERSON** become Duch?" asks the psychologist.

*Life events from a person's early childhood, education, and family aren't enough to explain how he or she comes to commit crimes against humanity. Geopolitical clinical psychology takes into account the relationship in each of us between our personal histories and our collective ones. It takes into account the effects that political, economic, historical, and cultural factors have on the subject's personality, alongside events in the subject's personal life—those that occurred in early childhood and the role of the family; in this case, it also takes into account the role mentors play in Cambodian culture.*

Duch told the psychologists about three events that particularly affected him; they took place during that crucial period before he went into the *maquis*, the period that shaped his thinking. The first is an utterly banal event of the kind found in any romance novel: a story of thwarted love. The second later triggers a flood of sarcasm among the trial's participants and observers: toward the end of 1965, someone stole Duch's bicycle, preventing him from getting to class at a time when teaching meant everything to him.

Each tiny event can seem meaningful when you're desperately seeking an explanation. But just as the road leading to mass murder is, in many ways, an indeterminable one of historical accident, becoming a mass murderer is often the uncertain and contingent fate of ordinary men. Those of us who have also suffered a romantic letdown or have had a bicycle stolen know that these are setbacks that can be overcome; they are without lasting damage. Yet Duch, for reasons of his own, remembers these events with sharp and painful clarity.

The third event is easier to link to his crime: Sihanouk's police arrested ten of his friends, including one he considered a brother, on suspicion of subversive activities.

According to Duch, these three events, whether directly or indirectly, helped drive him to Marxism.

# CHAPTER 21

**T**HE DEFENSE FAILS TO SEE WHY THE PROSECUTOR'S OFFICE insists on hearing from witnesses from M-13," says François Roux, trying his luck.

It's true that the tribunal's mandate doesn't extend to events that took place before the Khmer Rouge took power on April 17, 1975, or after it fell on January 6, 1979. All international courts are thus constrained in space and time, their mandates limited only to certain crimes or certain groups of people. The court in Phnom Penh has nothing to say about the five years of war that preceded Pol Pot's victory; it refrains from passing judgment on that war's hundreds of thousands of bombs and hundreds of thousands of dead. Likewise, the ECCC must ignore the twenty years of war that followed the fall of the men in black, those two decades rife with hundreds of thousands of land mines and refugees; and the court must ignore the way the international community compromised itself when it continued to recognize the Khmer Rouge as Cambodia's legitimate government for ten years after the party's downfall—this even *after* Pol Pot's crimes had come to light. Behind every international tribunal's limited mandate lies a cold political calculation, one that is often the consequence of the great powers taking steps to avoid any chance of incriminating themselves.

On the other hand, putting limits on an international tribunal is a sound precaution since, from the moment of their inception, such courts tend to become focused on their own survival and rarely restrict their own work. The men and women of the international judiciary are neither heroes nor saints. Whether serious and principled or vile and dishonest, they're never disinterested parties.

Yet, despite the limited mandate of the Phnom Penh tribunal, the answer to Roux's question is quite simple.

"There is continuity between M-13 and S-21," says the prosecutor firmly.

When he first entered the *maquis*, Duch briefly found himself under the command of Chhay Kim Huor, the teacher who had initiated him into the Revolution. But as of May 1971, he came under the authority of Vorn Vet, one of the movement's principle architects.

Vorn Vet was then secretary in charge of what the Communist guerrillas called the "special zone." M-13 was the police headquarters of that special zone. Vorn Vet asked Duch to run it. M-13 was where they developed those methods that, at S-21, were refined and practiced on a much greater scale.

The Khmer Rouge began its killings from the moment it took control of a tiny piece of territory. The most popular song in the *maquis* during those early days, says Duch, was called something like "The Cunning Infiltrator." Finding and annihilating the enemy within was explicit Party policy from the very start. The atmosphere of paranoia and terror, though of lesser magnitude than it would be in the coming years, was there from the beginning. Prisoners were called "spies" and declared guilty a priori. They deserved to die. There was no end to the executions. If higher-level officers were arrested, their subordinates had to be, too. No quarter was given to the "enemy," and that enemy didn't necessarily wear the uniform of the opposing army. The part of Duch's story that takes place during those years in the *maquis* is hard to swallow for those who maintain that the Revolution lost its way only after the Khmer Rouge's victory. How many revolutions are needed? How many victories?

It took comrades Duch and Pon no time to learn how to be torturers; they took to it with talent and dedication.

"I had no capacity for critical thinking at the time. The only thing that stayed stuck in my mind was the fear of being removed," says

Duch. It's not clear whether by "removed" he means losing his job or his life—the English interpreter gives nothing more than the word "removed."

Duch asked no questions when he was given the job. "We intellectuals had to be strict. Yes, I authorized torture. And I went to the interrogations myself."

At M-13, Duch first received "spies" sent from the zone controlled by Lon Nol's army. For the most part, they were poor people who hadn't had time to get away, he concedes. The great purges hadn't yet begun. Eliminating the camp's own personnel wasn't yet on the agenda. At this point in time, the Vietnamese were allied with the Khmer Rouge in the anti-imperialist struggle, and so weren't yet targets.

"The most shocking thing was the purge of the base," Duch says, referring to the masses in whose name the teachers were leading the Revolution. "It hurts every time I think about it."

Every fortnight, Duch went to a self-criticism meeting. Each person had to forsake his or her personal opinions and adopt the position held by the group. Individual consciousness was to be erased. Duch was no longer a citizen—he was the collective, he was the Party, he was the Revolution.

"In the psychology of extreme situations, the greatest danger occurs when an individual's affiliations are limited to one group," says the psychologist. "To safeguard against sacrificing individuality and self-awareness, it is vital to belong to more than one membership group."


JUST AS S-21 WAS COMPOSED of S-21 and S-24, M-13 was divided into two camps. One of Duch's deputies ran the first, where prisoners were "reeducated" and eventually freed. Duch himself ran the other, where people were held, interrogated, and, in all probability, killed. Duch remembers the names of some prisoners he was unable to free because, he says, the military chief, Ta Mok, opposed it. He freed around a dozen people in all, and once again, he remembers most of their names. "It was a very small number compared to the number of those tortured and killed. So it wasn't an act of valor. I cannot congratulate myself for it, but it's the truth. It is just a drop of water in the ocean of crimes I committed."

Nobody in that drop of water is more important to Duch than François Bizot.

Bizot was a twenty-five-year-old ethnologist when he arrived in Cambodia in 1966. He was conducting research at L'École Française d'Extrême-Orient, the French School of the Far East. For a century, this institution has been at the heart of the rediscovery and conservation of Angkor, the exceptional complex of temples in northern Cambodia. Bizot specializes in Khmer Buddhism. On October 10, 1971, while visiting a monastery, Bizot and his two Cambodian colleagues, Lay and Son, were ambushed by the Khmer Rouge. Bizot appeared before a summary people's court. He was accused of being a CIA agent and put through a mock execution. Then he was taken on a long march to a prison camp, where he was reunited with Lay and Son. According to Bizot, the officer who received him immediately showed himself to be "cynical and aggressive." Bizot was shackled to a metal rod to which ten or fifteen other prisoners were already chained. He had been walking for two days and two nights without washing. He was covered in mud. The officer denied him permission to wash. Bizot begged; a younger man intervened and told him, "Go and wash." It was Duch. Bizot had just arrived at M-13.

The camp was made up of three rudimentary, raised-floor shelters, one of which was reserved for sick prisoners. The captives had to urinate in a bamboo stick. To defecate, they had to go to a pit 1.5 meters wide, filled with excrement, a ditch which every prisoner "talked about with horror." Food was distributed twice a day. It consisted of "succulent rice, milled that morning by two prisoners." But there was nothing else with it. Malaria wreaked havoc.

"I was struck by Duch's poor health; like most of the prisoners he was unwell," explains Bizot. There was no medicine. Many of the prisoners who weren't executed died from their illnesses.

American B-52s were dropping bombs, says Duch, so they dug three trenches and kept the prisoners in them. The only thing the Khmer Rouge protected its prisoners from was bombs, lest they deny the guards the choice of when to kill them. There were ditches where prisoners were held and others where they were killed by a single blow to the back of the neck, economically, without making a sound.

Duch saw the B-52s passing high overhead, but he never experienced a bombing raid himself. He was never a combat cadre; he was a commissioner in the political police.

There were still neither prisoner lists nor archives. It was wartime. When a person was executed, his documents were destroyed with him. Over time, the cadres' diet deteriorated. The prisoners' diet, meanwhile, became downright inadequate. By the end of 1974, they were getting nothing more than rice dust.

M-13 was relocated three times in four years. When the camp was near a muddy river, the prisoners could wash and relieve themselves daily. The women were allowed to bathe unencumbered; the men were tied to each other with hammock rope.

According to Duch, the Khmer Rouge was still sparing children at that time. But this is ambiguous, to say the least. He describes how he looked after three children who somehow ended up at the camp. All three died. "They were with their parents at night, with me during the day. My superiors questioned my attitude, and I couldn't argue with them. They believed the children would avenge their parents. The three children died of illness. We let them swell up until they died."

M-13 was a prototype, one with notable shortcomings. For instance, one day in 1973, one of the prisoners managed to grab a guard's weapon. Mam Nai was wounded by another guard while chasing the fugitive. Like so much else, Mam Nai doesn't remember the incident. He says he was planting potatoes at the time. Some thirty prisoners escaped that day, leaving the camp practically empty. Duch was in the hot seat. "I told Vorn Vet to punish me. But he just sent me more people to destroy. It was probably my destiny to do that work."

"Did you consider doing something else? Did you contemplate escaping?"

"I never imagined anything but obeying orders to survive. I knew that my job was inherently criminal, but I had to follow orders. If that was my destiny, the one I couldn't avoid, then I had to carry out the tasks assigned to me."

Duch ordered executions. He remembers a few of his victims from those early days: a writer; one of Ta Mok's subordinates; cadres from Hanoi suspected of being Vietnamese spies. "My aim was to liberate my people, yet I did the opposite and became a part of the killing machine. M-13 wasn't only hard. It was cruel and odious. It was a place where we crushed humanity. It was beyond hard, beyond cruel."

Once, during the rainy season, around September 1974, the M-13 prison camp became heavily flooded. By around eight that morning, the water had risen significantly in the space of one hour. The prisoners were trapped, chained in the trenches. Duch claims they didn't drown, that everyone was moved to higher ground, and that the prisoners only died later, from illness. He says this emphatically and turns toward the public gallery, as though he's looking for someone in particular. He freezes, his mouth slightly open. "We didn't eat that day. Everything was floating around us."

His deputy Mam Nai remembers the flood. However, he cannot remember the fate of the prisoners. "I don't know if anyone died in the flood. I have no idea. Even pigs died," he says.

It wasn't unusual for Khmer Rouge leaders to concern themselves with the fate of animals. When Brother Number Two surrendered at the end of 1998, he said: "We regret not only the people but also the animals that lost their lives to the war. We are very sorry."

"I HATED POLICE WORK and I hated the killings," says Duch. "But they told me it was because of a lack of direction in the Party. At M-13, I came to hate shit, but I had to walk through it every day. I tried to solve things by my own means."

Duch turned to an old trick common to some Communist regimes since Stalin: the confession. He admits that extracting these sometimes got physical, for instance the day he was interrogating a prisoner while fighting a fever; two guards armed with pistols started to beat the prisoner, who gave in and made his "confession." Duch was infuriated not by the beating but by the fact that the prisoner failed to confess *before* it began. So, he says, he started hitting him himself, to punish him for not confessing until being beaten.

Duch says he used another prisoner, a poet, to test the techniques he had learned. The poet's interrogation lasted a month and Duch admits to beating him. Thus, Duch accounts for "at least two people" he admits to having beaten. "I don't remember the others now."

"So there could have been more than two and you don't remember the others?" asks the prosecutor.

"Yes. That's correct."

One witness says that he saw Duch whip a woman unconscious and laugh when she came to. But Duch vehemently denies this: "I interrogated that woman. I never beat a female prisoner. When a prisoner was beaten, no one helped me. The interrogations took place in the bush, away from everyone. I never let a prisoner see an interrogation. No one could see the interrogations."

Vorn Vet recommended using plastic bags. Duch, ever practical, was disinclined to do so—it wasn't easy to find plastic bags back then. The prisoners were often left tied to posts. "That's how I remember it," says Duch. The question arises of whether they were tied or suspended, as some witnesses have claimed. Duch gets to his feet and mimics the way a person was attached to a post. He uses the headphone wire of his simultaneous-interpreting device to show how the prisoners had their wrists and forearms tied. The prisoners sometimes stayed like that for four days. Someone asks if they were fed.

"I've forgotten."

One prisoner was burned with a torch. At least one other was made to stand in the cold wind. Women suffered the same treatment. But the young torturer declared this method inefficient. Furthermore, it offended Duch's sense of morality.

> When a woman's clothes cling to her body, you can see her shape and then suddenly there's a discomfort. Comrade Pon and I felt that discomfort. That's why we stopped when we did. Also, it was a useless method. The woman, whose name was Sok, didn't change her answer. She said that she had been sent alone, with no one to accompany her. I concluded that that type of torture was not only dangerous but could lead to an incident.

Duch remembers this woman in startling detail. She doesn't exist in any archive, and the events he is remembering go back thirty-eight years, to August 1971, but his memory is sharp.

"Her background was prostitution. She had been sent to spy in the liberated zone. That was what she said in her confession," he says, giving the impression of still believing it. "I asked her how old she was. She said twenty-eight years old. I told her to open her mouth so I could count her teeth. She didn't have thirty-two teeth. Anyone in their twenties has thirty-two teeth. So she was lying."

**NOBODY KNOWS HOW MANY DIED** at M-13. One former prisoner, who later became an auxiliary guard in the camp and who has since died, gave testimony to DC-Cam. The witness made a number of haphazard claims, including that thirty thousand people died at M-13, based on the following simple calculation: twenty executions a day over nearly four years add up to thirty thousand. The court has prudently decided not to use testimonies from such fickle sources as provided by NGOs.

According to Duch, between seventeen and twenty people worked at M-13, and there were never more than sixty prisoners there at any one time—both details match testimonies from others, including Bizot. The defendant claims that roughly two to three hundred people were executed at M-13 over those four years. Other experts, aware that the higher figure isn't credible but mindful of validating the executioner's own, succumb to a temptation commonly faced by those who tally the numbers in crimes against humanity: to split the difference and pick a number between the two extremes. People started saying that around three thousand victims died at M-13, but this is pure speculation. The reality, unpleasant as it may be, is that we have no idea. The remaining traces of the camp provide nothing on which to base any estimate; no document from M-13 survives; the Party had yet to develop its obsession with record-keeping, or the appetite for bureaucracy typical of totalitarian regimes. The Party was still at war and not yet in power. Duch says, "There were no orders to keep this sort of information. There was no reason to keep records about a person once he or she had been destroyed. The task had been accomplished. The mission was over."

**DUCH REMINDS THE COURT** that a twelve-year-old child is an adolescent. At twelve, a child can serve as a messenger. At sixteen, he can belong to a special execution unit.

> *I wasn't a role model when it came to killing, since I was frightened of doing it. We intellectuals assigned such tasks to children or peasants, who did them better. It was the same with Communist revolutionaries everywhere. The sons of the poorest families learned very quickly. I was from the intellectual class. I reminded them not to let people escape and to stop them from screaming or shouting. I never went there myself. I committed an enormous crime against the people of Amleang. They sent their children to me for education, so that they would become perfectly loyal to the Party. I say again, I hate shit and yet I walked through it. We had a good relationship with the villagers. They supported the Revolution. They agreed to send us their children to*

*help us. But they had no other choice. I felt a lot of affection for them.*
*I wanted to educate them so that they would join the revolutionary*
*path. The role of Party cadres was to train people to have a position on*
*class, to hold an absolute position against the enemy. But in reality, we*
*indoctrinated them to commit crimes.*

The M-13 prison camp was shut down after the Khmer Rouge victory in April 1975. We know that some prisoners were transferred to a new camp named, according to the same mathematical model, M-99. As for the personnel, a dozen or so interrogators, guards, and messengers quickly found new jobs at S-21.

# CHAPTER 22

*He was twenty-seven years old and I was thirty. I was so furious
at being mistaken for what I wasn't, for being accused of being a
CIA spy when such things were not even on my mind, that when he
questioned me I retaliated and asked my own questions back. This
went on for weeks and weeks.*

After he arrived at M-13 in October 1971, ethnographer François
Bizot was chained apart from the other prisoners, to one of the thin
pillars holding up a bamboo awning. The day after his arrival, Duch
began his interrogation. Bizot had to compose the first of a series of
"declarations of innocence." Duch wrote very late at night and very
early in the morning. He was known as a tireless worker who said little
and who took his responsibilities as camp commander very seriously.
Unlike the other Khmer Rouge men, says Bizot, Duch responded
whenever a prisoner greeted him.

For the prisoners, making confessions was an ordeal. But confessing was a way of life for everyone in the camp. The guards came
together for self-criticism sessions. First, each person took turns la-
menting his own revolutionary failures; then, he helped his neighbor
recall the mistakes he had made but no longer remembered.

"Was this to encourage people to denounce one another?" asks
Judge Lavergne.

*Absolutely. But, Your Honor, informing on others was considered
a good thing. It was a prerequisite, even. They held up as examples*

*those young revolutionaries who inculpated their parents without thinking twice. Denunciation, which is just another form of lying, is the very essence of the work of—how can I put it?—of spreading the Revolution.*

Bizot neither saw nor heard any violent acts during his captivity. His two friends told him that prisoners were beaten in the ribs with canes, but no one could see the marks under their black button-up shirts. After many days spent watching the guards and listening to them speaking among themselves, Bizot knew that they were beating prisoners. One day, during his daily swim in the river, he slipped away to the other bank. There he discovered a cabin, where he found "a vertical bar of thick bamboo with rattan rings attached to it, which," he realized, "were designed for tying wrists." On another occasion, he came across a former prisoner who had hung around there and who was busy whittling a rattan cane. Bizot called out to him: " 'Hey, comrade! Who are you going to beat with that rattan cane?' The poor guy looked at me and said, 'I'm not going to be *giving* the beating!' "

Duch toyed with his French prisoner at least twice. On the day he came to tell Bizot that he was going to be released, speaking French to him for the first time, Duch told his captive that he had been unmasked. Bizot fell to his knees, and then Duch said he was joking. On another occasion, aware of Bizot's friendship with Lay, Duch told him that he had to choose which of his Cambodian friends would go free and which would remain in captivity.

Yet despite this, once he was convinced of Bizot's innocence, Duch took a rare risk with his superiors. He went against the rule, and in front of the dreaded Ta Mok he asked for permission to release Bizot.

*I informed Vorn Vet that Bizot wasn't CIA. My superior laughed. He asked why I was frightened of the French. Ta Mok sent me a message saying, "Duch, never agree to release this researcher and the other two." I didn't reply. Vorn Vet came and I spoke with him. He went to find Ta Mok, who was eating. And that's when he told me,*

*"You can release him." There was a Party meeting, chaired by Ta Mok, at the end of which we gave the Party pamphlet to François Bizot. But only Bizot was released. The two Cambodians remained prisoners at M-13.*

Bizot is the only person who can say, "Duch released me." Consequently, Duch protects Bizot in the way of someone nurturing the hope of redemption. When a daughter was born to Duch a few years later, while he was running S-21, he gave her his grandmother's name—just as Bizot had done with his own daughter, Hélène. So, when Duch's "friend" Bizot steps onto the stand to testify, albeit indirectly, to the fact that the defendant personally inflicted violence on prisoners, Duch faces a real problem: how to protect both Bizot and himself. Someone asks him, "Who is telling the truth?" It's one of the rare instances during the trial when Duch does a miserable job of defending himself. He pretends to have never read Bizot's book, *The Gate*, only to quote from memory a passage "on page 169" a minute later.

**THE ETHNOLOGIST LOOKS DOWN** when he talks and keeps his body very still; he weighs his words, thinks hard, and gives testimony as though from a meditative state. Every now and then, a memory will bring a bitter smile to his lips; then he lowers his head again as a severe expression of concentration and hidden torment returns to his face. When he describes those terrible chains, his weariness and inner rage are obvious.

Bizot was scheduled to be released on Christmas Day, 1971. Once removed from his irons, he arranged for his two companions to be unshackled as well, though this only lasted for a short time.

*Needless to say, it was an incredibly significant reunion, but we didn't show it. The first thing we did was get together, though we didn't say much. For me, it was a reason to have hope. But it certainly wasn't for Lay and Son. They thought it was a way to make us swallow the*

*bitter pill, and neither they nor any one of my fellow prisoners believed
that they were actually letting me go. They all secretly thought that the
path I was going to take—the path they were leading me down—was
the same one my predecessors went down. Lies were the oxygen we
breathed and which we exhaled from our chests. They lied whether
they told someone he was going to be released or when they led him to
his death. They never told him what was really happening; they denied
everything until the very last moment.*

According to the defendant, Bizot's two colleagues were executed
about a year later in another camp.

"I believe I benefited from the presence of Lay and Son, but it
causes me pain," says Bizot while Duch listens, perfectly still.

During a recess, Bizot moves toward Duch a little, briefcase in
hand. They greet each other from a distance. The two men seem both
beholden to one another and crushed that they can't forget one another. Forty years after Duch decided Bizot's fate by pleading on his
behalf, Bizot's testimony will help decide the fate of his former jailer.
The day he received the summons to testify, Bizot felt a chill run
down his back. Yet by the next day, he was calm again. He was ready.

Bizot points out how during the two months of interrogations,
Duch made the kind of "human connection" with his prisoner that
no executioner should allow himself to make if he wants to put the
prisoner to death without scruple. Duch wouldn't make the same
mistake again. Bizot is convinced that if he hadn't been able to speak
Khmer, he would be dead. Not only that—if he hadn't had a strong
grasp of how to think and communicate in Khmer, he never would
have been able to establish a relationship with the camp commander.
Knowing a language doesn't mean knowing how to speak it, remarks
Bizot; it means knowing how to communicate in it.

Finally, it's highly likely that, had Bizot been captured a little
later, in 1972 or after, he would have found himself in very different
circumstances—M-13 would have evolved, and Duch would have been
far less gentle. In short, Bizot was a rare, youthful blunder in Duch's

career. Most of us are embarrassed when our past transgressions catch up with us. But Duch's blunder turned out to be a life-saver.

Bizot's release was pushed up a day. That evening, his ankles free of shackles, he approached Duch, who was standing by a fire. The two men had an almost normal conversation, chatting about their families. Duch was still single. Bizot was a young father. "Duch asked what had become of Hélène, my little daughter who had been with me in the car but who had remained in the last village, near the monastery. He tried to reassure me about her."

It was Christmas Eve. Bizot asked Duch who beat the prisoners.

> Duch told me right away that he sometimes beat the prisoners
> when they lied or when there were contradictions in their depositions;
> he said that he couldn't stand lies, and that this work—I forget the exact
> words he used—"made him nauseous," but that it was his responsibility.
> It's what the Angkar expected of him. This work was his responsibility.

The terror that took hold of Bizot when Duch admitted that he beat recalcitrant "spies" was so great it changed Bizot forever.

> Your Honor, I should say that until then, I had felt reassured.
> I believed that we were—that I was—on the right side of humanity;
> that some men were monsters and, thank heaven, I could never be
> one of them. I believed that this was a state of nature, that some of us
> were born evil while the rest of us could never be. But that evening,
> Duch's response, combined with my perception of him throughout
> the course of various interrogation episodes, opened my eyes. That
> Christmas Eve, I had expected to encounter someone inhumane, as we
> are accustomed to think of such monsters. But I realized that this was
> far more tragic and infinitely more terrifying, because in front of me
> stood a man who looked like many friends of mine: a Marxist who was
> prepared to die for his country and for the Revolution. The ultimate
> goal, for him, was the welfare and well-being of the inhabitants of
> Cambodia; he was fighting against injustice and inequity. And even

*if there was something insidious in the naiveté of the typical Khmer*
*peasant, there was also a fundamental sincerity in his beliefs, as is the*
*case with many revolutionaries. I myself had many friends in Paris at*
*the time who were committed to this Communist revolution and they*
*were looking at events in Cambodia with an outlook that, to me, was*
*horrifying. But in their eyes, the ends justified the means—in this case,*
*Cambodia's independence, Cambodia's right to self-determination,*
*the end to its citizens' misery, and so on. The Cambodians are not*
*the only people that have killed in the name of fulfilling a dream. So*
*here I was, looking for the first time behind the mask of the monster in*
*front of me. His job was to write up reports on the people sent to him*
*to be executed, and I saw that this monster was, in fact, human, which*
*was just as disturbing and terrifying. I was no longer sheltered from*
*this knowledge—we are no longer sheltered—and the worst mistake*
*we could make would be to separate such monsters into a different*
*category of being.*

Bizot's realization wasn't an instant one. It was the fruit of a long
and silent thought process that reached its culmination, and found
voice, the day Duch was found in 1999, more than twenty-seven years
after the ethnologist was released from M-13. A year after Duch was
arrested, François Bizot published *The Gate*, in which he told for the
first time the story of his captivity and what it taught him about humanity.

*I thought that if there were something to be said, it was that I had*
*known this man when he was a young man, a young revolutionary*
*who had been entrusted by his comrades with a particular mission*
*and who had done his job in a frightening but extremely rigorous*
*and thorough way, and always with a view to doing his job well and*
*fully. I then figured that it was necessary to make known that this*
*kind of tragedy was not committed by a monster, a different category*
*of being, but that this person was a human being like any other.*
*Consequently, it is necessary to distinguish what humans* do *from*

*what humans* are. *And I also realized that to be guilty of what one has* done *does not necessarily define what one* is. *Duch's situation did not allow him a way out, not just because he feared death—which was certainly a legitimate fear—but also because, with others watching, with the commitments he made by going into the* maquis, *he was part of a group, a kind of family, and it's very difficult to get out. The trap closed around him, and that continues to terrify me to this day, Your Honor.*

Early during his captivity at M-13, Bizot managed to get permission to keep a razor and a notebook in which he wrote a few childhood memories, a few poems, and some thoughts on Khmer Buddhism that might persuade his jailer that he was indeed a researcher and not a spy. Duch let him take the notebook with him after his release. Now, in court, Bizot pulls from his briefcase that curious, fragile relic, a sort of schoolchild's workbook with a big eagle printed on the cover. Suddenly seeing the notebook seems to unsettle Duch; he did, after all, read it closely some four decades ago. Bizot says that he has never managed to reread his prison journal. The few times he has opened it, he felt a great weariness come over him and had to stop. Of course, Bizot's journal contains no trace of the conversations he held with his captor, or any observations about the camp itself; for the prisoner to have committed such things to paper would have been a fatal mistake. Memory, therefore, is as much a reconstruction for Bizot as it is for the rest of us; he draws his memories through the prism of his "fears and emotions, of the things I felt at the time which have stayed with me for thirty-eight years." *The Gate* is of great value to history, to the human community, and to literature. But for the law, it's troubled waters. The cold but unavoidable burden of proof required by the law means that Bizot's extraordinary account is next to useless in court. The astonishing conversations that Bizot the prisoner held with his notorious jailer must, like so much else, be thrown out the window. Legally, they are worth next to nothing. Still, Bizot's testimony reflects part of the truth and elevates the trial: amid all the sordid facts,

Bizot's words reveal our tormented conscience; we hear it hammering as sharply as a woodpecker on an oak.

> *I must say that my encounter with Duch has left a mark on my destiny and determined everything I am today for a very simple and tragic reason: I must come to terms with a double reality—that of a man who was a vector, a tool of state-sanctioned killing, and I cannot imagine being in his shoes today with so much horror behind me. On the other hand, there is the recollection that I have of a young man who committed his life and his existence to a cause and to a purpose that was based on the idea that crime was not only legitimate but necessary. I don't know what to make of this, Your Honor. My experience brought me into intimate contact with this person and I cannot get rid of the idea that what Duch perpetrated could also have been perpetrated by someone else. By trying to understand this, I'm not trying at all to minimize it, or to minimize the reach and depth and horror of his crime. "His crime," that's where things get particularly difficult for me. I feel that these crimes were the crimes of a man, and that in order to understand their horror, we mustn't transform Duch into some kind of monster, but rather acknowledge his humanity, which is just like ours, and which obviously was not an obstacle, unfortunately, to the massive killings that were perpetrated. I fear that we have a far more terrifying understanding of the executioner when we measure him in human terms. And it is this awareness of the ambiguity of humanity that is the cause of my personal ambivalence today, Your Honor.*

Bizot has likely never voiced his doubts nor elucidated his suffering as clearly or as intensely as he does in court. It's only natural that his account of the facts should drift through the mirages of memory. In his writings, conversations, and solitude, Bizot had found avenues in which to express his pain, but it's only in the courtroom that he at last finds a way to convey its universal scope, and he does so brilliantly. He needed the solemn intensity of a courtroom to articulate it, he needed his daughter's presence, and he needed Duch's. Knowing that Duch was there, Bizot told me after the trial, helped him put his thoughts in order.

**THROUGH THE GLASS PARTITION,** Duch gives a wide and radiant smile to a tall and beautiful forty-year-old woman. It's the first time he has laid eyes on Hélène, who was three years old when her father was imprisoned. Bizot's daughter isn't spared the conflicting emotions tearing Bizot apart. Duch is the man from M-13 and S-21. And he's the man who saved her father's life.

I learned in Phnom Penh that many people don't like Bizot. He's not a man who takes precautions, and thus easily offends, rouses animosity, or generates antipathy. It's strange to hear him say that to know a language is to know how to communicate in it when he so often gives the impression of being blithely unconcerned by what others may feel.

One day, in the car taking us to the tribunal, he bluntly posed the following question to a Cambodian friend who had lost her parents under Pol Pot before moving to and growing up in the United States: "Would it shock you if I said that Duch is both the worst of men and a good man?"

A few long seconds pass.

"Yes. It still hurts," says the young woman without bristling.

"I understand," murmurs Bizot, with a sudden and touching note of affection and tenderness that he seems able to express only after he has given voice to his more pressing question.

The first time we lunched together one-on-one, we had hardly sat down when out of the blue he said: "Have you ever been betrayed?"

Like our Cambodian friend, I needed a few seconds to absorb the blow. I feigned reflection, then looked straight at him and said: "You mean, by a woman?"

In conversation, Bizot talks as though he's exempt from the usual conventions of politeness. You can react by either tensing up, replying in kind, and ending the conversation, or you can soften the blow by letting him occupy the position he's aiming for, appreciate the pain he's revealing, and agree to reflect on what it inspires. In court, Bizot avoids neither the ambiguity of man and our feelings, nor our thirst for vengeance. "The victims' screams must be heard, and we must never allow ourselves to think that their screams are too loud. The harshest condemnations we level against the defendant can never be harsh enough."

Bizot shows no hint of complacency in his testimony and no unnecessary harshness. Duch had expected the day of Bizot's testimony to be vital to his cause; instead, it's the day he realizes that his situation is beyond repair.

The survivor of M-13 lays his burden down before the judges. At the end of the afternoon, a much relieved François Bizot returns to where his daughter is staying in Phnom Penh. A woman is waiting for him in the street. She throws herself upon him, sobbing. She tells him that she watched his deposition on television. She is the sister of Son, one of Bizot's two murdered assistants. On October 10, 1971, little Hélène was at this woman's house when her father was captured. Bizot hasn't seen her since.

**THE DIFFICULTY WITH S-21,** writes David Chandler at the end of his book on the subject, "is not that it could be done, but that *we* could do it." Them. You. Me. If the historian's conclusion is so difficult to accept, perhaps it's because it stops us in our tracks: *In order to find the root of evil that was implemented every day at S-21, we should not look any further than ourselves.*

"In the culture of Democratic Kampuchea," Chandler explains to the court,

> the people who were given orders were accustomed to obeying. There is no questioning of authority in Cambodia. In a situation like S-21, obedience doesn't explain everything, but it's useful to see to what extent people like us can build a system where if the man in charge

*says it's okay, well, it's okay. Our capacity to commit evil is greater than our capacity to commit good. But that does not excuse people who kill. But I don't like people saying, 'Look at those evil people over there! We wouldn't—oh no, we would never do that, ever.' I don't want to say that what was happening at S-21 was done by another kind of people operating far away, but I want to suggest that under certain conditions—conditions that have happily been nonexistent in my own life—almost anyone could be led to commit acts like these. There is a dark side to all of us.*

The septuagenarian historian lets a moment of silence pass. *"C'est tout,"* he says in French, with a sad smile.

# CHAPTER 23

**P**HNOM PENH COMMEMORATES ITS LIBERATION TWICE A YEAR, though on neither date was the city actually liberated. On April 17, Phnom Penhers remember the city's liberation from the yokes of imperialism, feudalism, and military dictatorship: it was on this day in 1975 that Cambodia's Communists entered the city almost unopposed. Alas, that liberation quickly led to a nation of slaves and martyrs. Barely had Phnom Penh been liberated when she was begging to be liberated from her liberators, which is what her citizens celebrate on January 7: the day the Vietnamese army entered the city in 1979, meeting little resistance as they freed Phnom Penh from Pol Pot's terror. The price of that liberation was occupation by a much-loathed neighbor, one that many Cambodians have long suspected of wanting to annex their nation. So Phnom Penh was liberated, but still not free. Ten years later, on September 26, 1989, the Vietnamese officially withdrew from the capital, and Phnom Penh was liberated for the third time in a decade and a half. This third liberation is the only one that didn't happen by force, and the only one that didn't immediately create the need to rid the city of its liberators. It's the only one that isn't commemorated today.

Phnom Penh is a city that gives herself without reservation, that lets herself be abused without flinching. She abandons herself to those who possess her just as she does to those who inhabit her, and she regrets it too late. She is a small town grown big, still curiously sweet and kind despite the devastations she has suffered. Discreet, vulnerable, radiant, Phnom Penh is a city that, despite her haunting past, continues to embody her citizens' finest qualities.

**IT WAS THE VIETNAMESE ARMY** that, during the second liberation of Phnom Penh, discovered first S-21 and then the killing field at Choeung Ek. Both sites quickly became the principal memorials to the crimes committed in Cambodia between 1975 and 1979. S-21 was rapidly turned into a "genocide museum," while a massive Buddhist memorial, or *stupa*, consisting of hundreds of skulls sitting on shelves behind glass in a column, was raised on the edge of the graves of Choeung Ek.

Today, at S-21, in the two square courtyards on either side of Building E, in which the painters Bou Meng and Vann Nath were locked up and ordered to paint Brother Number One, there are paths crisscrossing four rectangular lawns lined with frangipani, palm, and mango trees. I was resting on a bench there one day when three Cambodian students from the Royal University approached me and asked me to participate in a survey. Their question was: should the nation continue to use the U.S. dollar alongside its own currency, or should the riel be the only legal tender?

Disconcerted, I stammered: "Isn't it a bit peculiar to carry out this study here?"

"Yes, it is," said one of the students, impeccably polite. "But we chose it because you meet foreigners here."

They're not wrong. Nowadays, most tourists who visit the Cambodian capital visit four places: the Royal Palace and the National Museum on the one hand, and S-21 and Choeung Ek on the other. In Cambodia, perhaps more than anywhere else, mass tourism has taken on mass murder.

Paradoxically, the place is a bit of a mess. The level of maintenance at S-21 is perfunctory at best, which gives it a crude, unvarnished atmosphere. Part of the brutality of the experience of visiting S-21, of feeling it in your body, comes from the stains on the walls and floors and the rings caused by moisture; from the dusty, forgotten corners underneath the staircases cluttered with old signs, liquor bottles, barbed wire, and worn boards; from the relics of previous exhibitions now falling into decay, such as the plaster puppet that once hung from the portico in the courtyard; and from the objects left over from the crime itself, such as the old, accordion-bellows camera on its tripod, or the two boxes filled with rags, bones, and skulls.

S-21 draws much of its grim power from its shabby maintenance, even if it is a little unseemly at times. The ceilings in Building A, where the most important prisoners were kept, are dangerously close to collapse, and have been hastily propped up with provisional wooden shoring. In Building B, the ceilings are actually collapsing. A deep crack spans the load-bearing wall beneath the main staircase in Building C. In mid-2010, the museum's director told me he could foresee one day having to close the old prison cells to the public to avoid accidents. Rubbish, piles of broken bricks, the carcasses of various objects, and wild banana trees litter the empty lots behind the buildings.

The former cells and interrogation rooms, their yellow-and-white-tiled floors turning brown and their distempered walls scarred with haunting abrasions, feel frozen in time. In these rooms, where torture was carried out until the very last day, large photographs of the swollen bodies found tied to metal bed frames are so faded that their horror seems to have surrendered to the modesty instilled by the passage of time. The buildings' façades, shutters, and balconies have all acquired the weathered, dusty colors of disuse. The clouds of bats that have colonized the complex sleep peacefully under its more remote roofs, and splatter the ground beneath them with their droppings. Nothing is really filthy and yet everywhere feels dirty.

One room, kept in particularly good condition, is forbidden to visitors and permanently air-conditioned. It is the room where the incalculably valuable S-21 archives are kept: 6,147 photos, of which 5,382 are of prisoners; 4,186 write-ups of interrogations; 6,226 biographies. Duch's treasure trove.

The precious confessions are filed in dozens of black boxes lined up on the shelves of five large, old, wooden bookcases. Little seems to have changed in the layout of the archive room in at least fifteen years, and there's something particularly admirable about the staff preserving it in this way. On the day I visited with a journalist friend, I watched one of the archivists patiently remove the rusted old staples holding the sheets of paper together. Her attention to the fragility of the archives was moving: first, she covered the staple holes with a small square of adhesive tape; then, she replaced the staples with plastic paperclips, making sure to slip a small piece of paper between the file and the clip. The fragility of those documents, as thin as Bible paper, sent a chill down my spine. On the interrogation summary she was holding, I read: NEOU PHEAP, 27/4/76, SEVENTEEN YEARS OLD.

These confessions were the core of the concentration-camp machine. Yet none are publicly displayed and their importance isn't explained to visitors, who never get to see the meticulous margin notes made by Duch, his superiors, or his subordinates. There's no public exhibit of what produced this hellhole before it destroyed people, nor does the public get to see the death lists handed to the drivers who transported the victims to Choeung Ek. The museum draws its strength from the fact that it offers visitors not just sights but feelings. What's on display isn't necessarily a representation of what it was like, and what it was like isn't clearly explained.

In one of the interior courtyards are some "educational" information panels written in the language of pro-Vietnamese Communist propaganda, strewn with dated and embarrassing turns of phrase denouncing the "Pol Pot–Ieng Sary clique." No tourist can resist taking photos of the famous text, writ large on one of the panels, of the ten commandments regulating the prison's "security agents" (the French version is hardly less crude than the English):

1. You must answer accordingly to my question. Don't turn them away.

2. Don't try to hide the facts by making pretexts this and that. You are strictly prohibited to contest me.
3. Don't be fool for you are a chap who dare to thwart the revolution.
4. You must immediately answer my questions without wasting time to reflect.
5. Don't tell me either about your immoralities or the essence of the revolution.
6. While getting lashes or electrification you must not cry at all.
7. Do nothing, sit still and wait for my orders. If there is no order, keep quiet. When I ask you to do something, you must do it right away without protesting.
8. Don't make pretext about Kampuchea Krom in order to hide your secret or traitor.*
9. If you don't follow all the above rules, you shall get many lashes of electric wire.
10. If you disobey any point of my regulations you shall get either ten lashes or five shocks of electric discharge.

Yet these commandments didn't actually exist, at least not in the form in which they're presented: no instance of them can be found in S-21's voluminous archives. This text, which makes such an impression on visitors, and in front of which tourists never fail to cluster, is a reconstitution based on a number of testimonies. The rules are similar to those imposed on the victims of S-21. But nonetheless they constitute a false truth in a place where, at its height, a false confes-

---

* Kampuchea Krom (literally, "Cambodia from below"), encompassing the Mekong Delta (formerly Cochinchina), was part of the Kingdom of Cambodia before the Vietnamese annexed it during the eighteenth and nineteenth centuries, and was granted to Vietnam under French colonization. The Khmer Rouge suspected the Khmer Krom of spying for the Vietnamese.

sion meant a death sentence. The construction of collective memory doesn't trouble itself with scruples.

Likewise, the torture instruments on show were collected during the Vietnamese occupation, and while some of the objects and tools are from S-21, others are not. A chilling rectangular wooden bathtub, for example, to the bottom of which prisoners were attached by the feet, was recovered from elsewhere.

In the exhibition rooms, photos drawn from different archives have long been juxtaposed in the greatest possible disorder, devoid of captions or any chronological or thematic arrangement. For years, visitors have been viewing them without the slightest idea of what they were looking at. Lost in the middle of it all is a photo taken in 1981 of the seven people then known to have survived S-21. Among them are the painters Vann Nath and Bou Meng, as well as the mechanic Chum Mey, who continues to come to the museum three times a week to earn a small income as a guide. Some tourists, if they are accompanied by an interpreter, are privileged enough to visit S-21 in the company of one of its three surviving ex-prisoners. One day, two months prior to the start of Duch's trial, I found Chum Mey standing in front of the glass board in which his uncaptioned photo sits next to equally anonymous photos of his torturers. The survivor with the lustrous white hair was trying to make three American tourists understand that that was him standing there in the photo; that he was Chum Mey. When their good fortune dawned on one of the Americans, he quickly made his friends stand next to Chum Mey and the thirty-year-old photo of him.

Say cheese. And then they left.

Under a staircase in Building C, in a corner hidden from view, there's a crumbling plaster wall on which tourists have marked their "thoughts," reminders of their important visit to S-21. All the absurdity of the modern tourist is revealed in their graffiti:

Life is what you make it.

Breathe and Smile.

One life & live it ☺

Jesus is our love.

Don't let shit like this ever happen again. Please!

Remember CIA interrogation at Abu Ghraib, IRAQ; don't be so fucking ignorant, they also torture. [This is followed by an e-mail address.]

At the corner inside S-21 where tourists reach the end of their visit, a souvenir shop sells "Pol Pot sandals"—simple soles cut from tires and fitted with black rubber straps—as well as sunglasses, silverware, stamps, counterfeit bills, and pirated copies of books and DVDs. Their connection with the history of the Khmer Rouge is weak, at best. But today's Cambodia, for better or worse, is open for business.

Two weeks after the verdict is handed down in Duch's trial, I spend an afternoon taking two friends visiting from overseas around S-21. In the courtyard, in the shade of the frangipani trees, their yellow-and-white flowers quietly beautiful despite their empty hearts, Bou Meng and Chum Mey sit on a bench, talking and waiting. Chum Mey is hoping to earn a few dollars from tourists. Bou Meng and his devoted young wife are planning to sell copies of the autobiography he has just written. Tourists walk past without paying them any attention. The two survivors watch, unable to communicate. Bou Meng is going on a trip to a Scandinavian country soon, and is pleased about it; he proudly shows me the little UNESCO badge he wears on his belt. The museum is now a World Heritage site. Major renovation work has begun. The main entrance is being moved, so tourists will no longer enter through the victims' gate. The roof of Building A is being restored. The rear of Buildings B and C has been cleared of the banana trees and rubbish.

Shops also fill the ground around the entrance through the perimeter fence protecting Choeung Ek. A journalist friend of mine described the skulls and skeletons in the memorial here as "simple evidence of a complicated horror." The thousands of bones found in the killing field are exhibited over seventeen floors in the sixty-two-meter-high glass-and-marble tower, and constitute a perfect visual feast for *homo touristicus*, who remembers not what he sees but what he photographs.

Thanatourism, or dark tourism, is already a mass phenomenon. It's also lucrative. In 2005, some Cambodians were shocked by two instances of it. First, a young man opened Café History opposite S-21, where waitstaff in black pajamas and red *krama*s around their necks (the uniform of the Khmer Rouge) offered tourists a set menu comprising a vile gray soup, an egg-based dessert, and tea. A Khmer Rouge lunch, in other words, all for just $6. The authorities were quick to shut down this genocide-tourism entrepreneur. Around the same time, Cambodians were finding out that the operating concession for Choeung Ek had been privatized and, moreover, granted to a mysterious foreign company called JC Royal, registered in Japan. But it turned out that the not-insubstantial revenue from Choeung Ek—$622,000 for the 2006–07 financial year—disappeared not into the hands of the mysterious Japanese outfit but into a supposedly not-for-profit fund with which several highly placed members of the government had very close ties.

Thirty years after 9,000 bodies were exhumed from Choeung Ek, the site serves virtually no educational function. Like at S-21, the explanatory panels put up at Choeung Ek in 1988 during the Vietnamese occupation remain in place today, and are read by the 200,000 tourists and some 20,000 Cambodians who visit the site annually. They inform us of "Pol Pot's gang of criminals" and those who "have the human form but whose hearts are demon's [sic] hearts, they have the Khmer face but their activities are purely reactionary."

Choeung Ek not only stirs up financial appetites, but also bitter political quarrels. While former Khmer Rouges in power today view April 17 as a day of liberation, others from the main opposition party think it marks the beginning of the nation's tragedy, and gather at Choeung Ek on that date not to celebrate the liberation but to remember the oppression that followed. Around a month later, on May 20, members of the ruling party also congregate at Choeung Ek to lament the tragic setbacks faced by the revolution that many of them served, including the state's three highest representatives who, if they hadn't had the presence of mind to flee the purges in 1977, most likely would have ended their days in this very field, with a quick blow to the back of the neck. So, on May 20, they organize an official "day of hate." A few actors dress up in the Khmer Rouge's black uniform and tie around their waists red-and-white *kramas*, the checkered cotton scarves so popular with Cambodian peasants, who use them as hats, bags, loincloths, and swaddling clothes. The actors enact torture scenes in which other actors kneel on the grass and, with terrified expressions, beg for mercy. Recently, after one such ceremony, the deputy governor of Phnom Penh said that its purpose was to "help us remember who saved us and who killed us."

**DUCH COMMITTED TWO FATAL ERRORS** when he left S-21: he failed to destroy the archives and he let a painter live. Not only did he leave thousands of pages that document his crime, he spared the artist who, with his brush, would prove the most devastating witness against Duch. The paintings Vann Nath made while imprisoned at S-21 have disappeared. But those he painted after the liberation have helped forge our image of the terror that reigned in that prison and the tortures inflicted there. No other testimony given over the past thirty years matches the power of the fourteen works by the artist-survivor. Visitors to the museum never fail to be struck by them.

But it is the photographs of the prisoners that anchor the experience of visiting S-21. Around two thousand portraits are exhibited in what were once classrooms, then prison cells, and now museum rooms. Their subjects look frightened, questioning, restless, quiet, defiant, smiling, tired, swollen, puffed up, gentle, jocular, determined, shocked, stiff, confident, obedient, despondent, resigned, evasive, astonished, sweet, sad, anxious, exhausted, proud. They are young, old, good-looking, ugly, baby-faced, thin, plump, blindfolded, and tied up. There is nothing more crushing than seeing these portraits hung tightly together, panel after panel, room after room. The intellectual power, emotional charge, documentary, and even artistic value of these snapshots of the thousands who died in the days or weeks after their photos were taken are what both define and anchor memory at S-21.

Yet almost everyone who visits S-21 walks past the photographs of its victims utterly unaware of the ambiguity inherent in many of them, including in the famous, harrowing image of a beautiful woman, understated and elegant, her hair slightly disheveled, her expression one of exhaustion, despair, and resignation, who is holding on her knees a sleeping infant in diapers, its eyes closed and hair slick with sweat. This woman, who was murdered in 1978 at S-21, was herself a revolutionary, the wife of the secretary of the southeast region, one of the regime's high officials who fell from grace and was eliminated, along with his family, by the regime he had served.

It is estimated that three-quarters of the victims at S-21 were themselves Khmer Rouge, of high and low rank, all of them destroyed by the regime.

"Every security-service post in the country, S-21 included, was tasked with imprisoning, interrogating, torturing, and, finally, smashing—that is to say, killing—people. But one principle unique to S-21 was that it was tasked with killing members of the Central Committee," says Duch.

Throughout the world, S-21 has become the symbol of the Khmer holocaust, its most famous memorial, an emblem of the massacre for tourists to visit and, now, its judicial epitaph. Yet the bulk of its victims—an estimated 80 percent—were themselves members of the Khmer Rouge. No doubt there were those who were under its thumb, but many gave themselves wholeheartedly to the regime, and would be in the dock today had they not been annihilated by their own party.

The ledger of S-21 dead provides an X-ray image of the Khmer Rouge's internal purges. "In single-party regimes, purges are a normal phenomenon, not unlike political crises in France," said Raymond Aron, with a mix of irony and seriousness. In mid-1977, Cambodia's central zone was purged. At the end of 1977 and beginning of 1978, it was the turn of the northern zone and then, in the second quarter of 1978, the eastern zone. More than a thousand Khmer Rouge cadres from the eastern and northern zones were sent to S-21.

Duch's trial is that of a Khmer Rouge cadre who killed primarily other Khmer Rouges. S-21 was the site of the regime's centralized purges rather than of mass murder on a national scale. For the millions of Cambodians who were annihilated by a Khmer Rouge regime they never served, there's a grim irony in this partial misunderstanding. S-21 was the most political of Democratic Kampuchea's two hundred documented security centers, and one could argue that the crimes committed there should be of lesser priority for the court, if you accept that many of the victims at S-21 had themselves been torturers or accomplices.

"Were any of them better than the others? Who *didn't* have blood on his hands?" says Duch about three victims of S-21 whose names regularly come up during the trial: a member of the Central Committee and the two teachers who introduced him to the Revolution.

Duch's trial has done justice to Vorn Vet, Duch's former boss at M-13 and a member of the standing committee, who was killed just before the Vietnamese arrived; Ban Sarin, one-time head of internal security in the northern zone, who was destroyed in January 1977 after having served the movement for fifteen years; Koy Thuon, former secretary of the northern zone, minister of commerce, who was promoted thanks to the purges of 1972 and who, according to Duch, had the authority to wipe out people before he was himself executed in April 1977; Ney Saran, alias "Ya," who became secretary of the northeastern zone after serving as Son Sen's deputy, and who was killed in October 1976; Sy, who was first Ta Mok's deputy and then secretary of the western zone, executed in April 1978; Ros Nhim, secretary of the northwestern zone, purged the following month; Nath, the former director of S-21; Nun Huy, the former head of S-24; and so on.

It again befalls Duch to articulate a troublesome fact that many would rather forget. "The life of a Central Committee member equals the lives of thousands of ordinary people. What do you say to that?"


**NOBODY DESERVED TO DIE** the kind of death meted out at S-21 and Choeung Ek. Though the two sites can bring out the worst kind of behavior in both visitors and carpetbaggers, S-21 and Choeung Ek remain memorials to the suffering inflicted by the Khmer Rouge and to the crimes they committed. This is not the case at Anlong Veng, the Khmer Rouge's final stronghold, where a more extreme version of dark tourism is taking shape.

# CHAPTER 24

**F**OR A LONG TIME, THE ROAD TO ANLONG VENG WAS SUFFI-
ciently rough and corrugated by rain to dissuade most visitors, including most tourists, from venturing there. The sleepy, remote
little town lies at the foot of the Dangrek Mountains on Cambodia's
northern border. Its location appears to serve a dual purpose: on the
one hand, it is protected from Thailand by the mountains, but on the
other hand, it provides easy refuge.

Just before the bridge leading into town, a track veers off to the
left. The track leads three hundred meters out onto a small peninsula
covered in mango trees and jutting out into an area that is both land
and water, known locally as "the lake." The subtle, muted combination of water, wild grasses, and tall, bare trees soaring spear-like into
the sky infuses the area with a meditative peace. If you stand on the
peninsula and look out over this marsh, you can see the village of
Anlong Veng without being seen from it. This promontory suits wise
men, thinkers, and soldiers on watch, and it's here that Ta Mok, the
most powerful and brutal of Khmer Rouge military commanders,
lived.

Anlong Veng was the last bastion of the Big Brothers of the Revolution before they all died or surrendered. It's where Pol Pot and his
most faithful associates lived out the last decade of the war, from
1988 until 1998. But Anlong Veng's true master was Ta Mok. Known
within the Politburo first as Brother Number Five, then, with each
successive purge or defection, as Brother Number Four, then Number Three, Ta Mok gained a reputation for being the most ruthless

member of a cohort in which competition was fierce. It is said that a small house once stood on one of the strips of land jutting out into the middle of the lake, opposite Ta Mok's house. That is where Pol Pot is said to have stayed whenever he came down from the Dangrek Mountains to pay a visit to the region's strongman. All that remains of it now is the outhouse.

In 1997, Ta Mok had Pol Pot arrested, summarily judged, and placed under house arrest. Pol Pot had just ordered the deaths of Duch's old boss Son Sen, Son's formidable wife, and eleven members of his family. "Paranoia moves at a gallop; it never stops. Nothing appeases the paranoid man," says the court psychologist.

For his own friends, Brother Number One was clearly becoming too dangerous. Only Ta Mok could take him down. Pol Pot died less than a year later, in April 1998. By the end of that year, Ta Mok was the only remaining Khmer Rouge leader not to have surrendered. He was eventually captured in early 1999. Ta Mok was the only Khmer Rouge leader—along with Duch, who was terrified of him—to be imprisoned. In July 2006, exactly ten days after the international tribunal with the jurisdiction to prosecute him was officially established, Ta Mok died, effectively thumbing his nose at humanity and its bourgeois system of justice one last time.

The first time I visited Anlong Veng, Nhem En was deputy district governor. Nhem En joined the victorious Khmer Rouge Army in 1975, at the age of sixteen. The conflict was still going on twenty years later, but by then Nhem En had realized that the army he'd joined, which was, once again, a guerrilla force, had crumbled. The Cold War was over. After ten years of occupation, the Vietnamese Army had gone home. In 1991, all of Cambodia's other political forces had agreed to a peace settlement; the following year, the UN's largest-ever peacekeeping operation was established in Cambodia. Elections were held in spite of the war. The Communists installed by the Vietnamese and loyal to Hun Sen now shared power with the royalists. Like many former Communists, they turned toward the most ferocious form of capitalism. Only the Khmer Rouge rejected both the peace process and the free-market economy, which in the eyes of many made them a tiresome and disconcerting anachronism. There was also, of course, the terrible burden of their blood-soaked past. Defections spiraled. It was time for the waning Revolution's cadres and soldiers to take advantage of the amnesty agreement signed by the king, and of the national reconciliation promised by Hun Sen. The Khmer Rouge was in its last throes, its old, paranoid heavyweights settling scores between themselves: Pol Pot murdered Son Sen, Ta Mok arrested Pol Pot, and so on. It was time to jump ship. Twenty years after embracing the Revolution, Nhem En bade it farewell.

While Duch slipped into another identity, Nhem En weaned himself of his revolutionary habits. But whereas Duch knew that his past at S-21 would prove fatal if discovered, Nhem En found that his past could bring him glory and, he hoped, income.

In the mid-1990s, S-21 was already famous throughout the world, not because of Duch but because of the photo portraits of its victims. Professionals studied and praised the particular artistic quality of those thousands of black-and-white images. It wasn't long before their creator was found: Nhem En, fresh from the jungle, was dubbed the "S-21 photographer." Soon he was decorated by the American ambassador and invited to New York; media outlets competed for interviews. Nhem En believes he granted between one and two hundred, for which he charged as much as he could get away with.

In the three years following his return, while giving all those interviews, Nhem En also worked for the ministry of the interior on the demobilization and reintegration of his former brothers-in-arms. But in 1998, the war was finally over and Nhem En found himself out of work. He left for the rural seat of Anlong Veng, where he joined the royalist party and became its local representative. In 2005, though he became deputy governor, he realized that the real power was no longer in the royalist party's hands. What's more, three years previously, one of his sons had been sentenced to eighteen years in prison for murdering his wife. Nhem En told me that he needed between $10,000 and $15,000 to pay off the judge and get his son out. If he wanted to help his son, and if he wanted to help himself, then his interest indubitably lay in joining the real winners of both the war and the post-war period: the Cambodian People's Party, led by Hun Sen, which has been in power since 1979. So, in 2006, in a move typical of Cambodia's pragmatic, fickle politicians, Nhem En switched sides again. He had been a member first of the Khmer Rouge, then of the royalist party. Now he is a member of the party that defeated, subdued, or absorbed them both.

A deputy governor earns a salary of $35 a month, at least in theory, Nhem En tells me. He lives in a modest house across from a school built by Ta Mok. The walls of the entrance hall are papered with photographs, including one of the American ambassador awarding him a prize for his photography at S-21, and with Khmer Rouge propaganda posters, whose proletarian realism seems long out of date. Nhem En has accumulated around two thousand photos linked to the Khmer Rouge, which he keeps in albums at his home. Most of them weren't taken by him; he collected them from other sources. Though many are of a very average quality, they often have a documentary value. These aren't photos of S-21, obviously—those were left behind when the staff fled the prison. What Nhem En has is a mixed bag of images, some well-known, others less so. Many are of bodies: the murdered Son Sen and his wife covered in their own blood, Pol Pot and Ta Mok lying peacefully on their deathbeds. Nhem En took the photos of Ta Mok. It becomes clear to anyone looking at these pictures that it's highly unlikely that Nhem En was ever the chief photographer at S-21.

Nhem En has tried to sell his albums, but in vain. He needs money. He would like to make more profit from his past at S-21, his biggest asset for the past ten years. He became aware of the media's interest in genocide very early on. He tasted a few of the rewards. Now he would like to develop what he sees as Anlong Veng's tourism potential. After all, three of the most famous and most cruel leaders of Democratic Kampuchea are buried there: Pol Pot, Ta Mok, and Son

Sen. It's the end of 2007, and Nhem En is planning to build a museum devoted to the Khmer Rouge.

It's hardly an outlandish idea. In December 2001, a government directive promised to "examine, restore, and preserve existing memorials, as well as investigate and study other remaining mass graves so that all such places may be transformed into memorials with fencing, trees, and information panels for both citizens and tourists." Local authorities intended to create a "national region of historical tourism" in the mountains of Anlong Veng, which had been the setting for the "final stage of the political lives of the Khmer Rouge leaders and military organization." Nhem En is only positioning himself in the new dark-tourism market, a market which, in Anlong Veng, is taking its most eccentric form: here, tourists are invited to make pilgrimages to the tombs and homes of Cambodia's mass murderers.

Nhem En thinks big. He calls for an investment of $2 million, including $300,000 for the museum. "It's a good project. It would be useful to generate money for my region," he tells the press. "I think that international tourists will want to see the portraits of the Khmer Rouge leaders. We also need further advice: should the museum be devoted only to the Khmer Rouge leaders, or should it include other things, too?"

Nhem En thinks Anlong Veng could become as well-known as certain places in Germany or Vietnam, because "Ho Chi Minh, Hitler, and Stalin are all heroes of something, whether good or bad," he says. He tells me that he's seeking both technical and spiritual support.

Above all, he's seeking money. He introduces me to his business partner, a wealthy jeweler from Siem Reap, gateway to the temples at Angkor and one of the country's most popular tourist destinations. Nhem En has convinced her to purchase a large parcel of land by the road leading into Anlong Veng. This is where he hopes to build his Khmer Rouge museum, using the hundreds of random photos in his possession as the basis of its collection.

The jeweler is a seductive, elegant, slightly eccentric divorcée.

She built for her children an enormous house in the shape of a cart drawn by two bulls. There's a confident and intelligent gleam in her eyes, and she clearly manages her business with authority and success. In any case, you can't lose money on property in Cambodia, where one of the main causes of social and political violence is the frenzied, brutal land rush taking place. In Cambodia today, people are no longer killed or deported for their ideology. Instead, people are violently expelled from and killed for their land. Some Khmers Rouges have become today's "Khmers Rich," their greed, racketeering, and corruption equal to the vices of the ruling class of the 1960s they so scorned.

Whatever becomes of the museum, it will only be a small part of the jeweler's portfolio; she says she wants to build it on only one hectare out of the parcel's fifty-five. She knows that the road between Siem Reap and Anlong Veng is to be rebuilt within two years, after which the journey linking the two will take only an hour and a half. The businesswoman believes that Anlong Veng could then be offered as an additional destination to the million tourists who visit Angkor each year.

"While visiting Angkor Wat, they could also learn what happened in Cambodia. I've discussed it with travel agents and they support us," she told me.

If tourists visiting Phnom Penh have made S-21 and the killing fields at Choeung Ek obligatory stops on their itineraries, why wouldn't at least some of the crowds that flock to the splendors of Angkor every year consider visiting the graves of Cambodia's mass murderers? Neither to revere nor to revile, but simply to see?

Nhem En has been thinking about his museum since 2000. On the day I visited him in November 2007, he told me that he had just signed a contract with the governor; he believed he was getting closer to his goal. But the beautiful jeweler and the twisted photographer were always an unlikely team—a few months later, she pulled out of the project. Nhem En again came into the spotlight when he announced that he was selling all his treasures for half a million dollars, including a pair of sandals which, he claims, belonged to Pol Pot, as well as a piece of the former tyrant's toilet.

**NHEM EN, LIKE EVERYONE ELSE,** does not speak the "whole truth and nothing but the truth." During those years when he passed himself off as the head photographer of S-21, he didn't mention that there had, in fact, been six photographers working at the prison and that, at seventeen, he hadn't been their boss. The story he gave in his paid interviews suited him; but after a while, the well began to run dry, and his credibility with it. It's likely that it will suffer even more with the trial, since two other photographers are still alive, and one of them has already been interviewed by the investigating judges.

One day, I found myself in a car with Nhem En at the wheel. While negotiating the ruts and potholes, he admitted that someone else had been the chief photographer at S-21. Speaking of Duch, he told me that "those who commit such crimes always have a reason to conceal the truth and to lie." There's no question he knows what he's talking about.

Each step of how S-21 operated is described during Duch's trial: how a person was arrested, registered, and locked up in a cell; how a person was interrogated and tortured until he confessed; how a person was taken to the killing fields. The only stage for which no former member of S-21 has been summoned is the one without which S-21 wouldn't be the world-famous museum it is today: the moment when the prisoner was photographed. Among the photographers who were interviewed during the trial's investigative phase, Nhem En was the only one whose deposition was read in court. He wasn't even called to testify. By now, few people still take him seriously.

# CHAPTER 25

**T**A MOK'S ABODE IN ANLONG VENG IS ALREADY BEING COMMER-cially exploited. Foreigners pay $2 a visit and 500 riels (about 12 cents) to use the toilets. According to the province's director of tourism, around a thousand Cambodians a month visited in 2007, along with fifty to a hundred Thais and twenty or so other foreigners, mostly Japanese. Out on the beautiful peninsula, in the garage where cars were once parked, two big cages resembling giant fish traps draw the interest of a group of Cambodian tourists. They are prisoner cages, someone explains a little too eagerly. Farther on, among the mango trees, lies the ruined chassis of a truck. With its hood propped open, it looks like some sort of a giant, yawning toad. It was once the mobile broadcasting truck used by the Khmer Rouge's television propaganda service.

The main house, though entirely emptied of its furnishings, has been pretty well-maintained. On the ground floor are murals painted in the early 1990s, illustrating the Khmer kingdom's former splendor—something that remains a source of pride for a patriotism with few sources from which to draw inspiration: there's the temple at Angkor Wat, of course, celebrated even by the Khmer Rouge, who, despite their determination to erase all traces of counterrevolutionary culture, placed an outline of the temple in the center of their flag; and there's the temple at Preah Vihear, a spectacular ruin perched atop a cliff in the Dangreks, which to this day continues to cost Cambodian and Thai soldiers their lives, ludicrously sacrificed on the altar of patriotic madness as both nations claim ownership of it. To sit on

Preah Vihear's rocky outcrop as dawn's red glow turns to silver is to witness the majestic morning of the world as it was before man, and as it remains despite him.

On one of the walls of Ta Mok's house, next to some paintings of the famous temples, there's a large map of the kingdom. Nhem En produced it in 1993 on behalf of general command, which is where it remained until it was brought here. The room gives onto a large terrace with a magnificent view of the lake. Four Cambodians are talking quietly on the terrace. Ta Mok lived in paradise, they say, admiring the landscape. All four have come for a wedding. They're delighted to be able to combine the wedding with two lifelong dreams.

"I always wanted to see Angkor Wat and Anlong Veng. I'm so happy my dream has come true," one of them tells me.

He says he'll pay a visit to Pol Pot's tomb after the wedding. Though disinclined to talk, the four tell me that they had all been about ten years old when the Khmer Rouge seized power, and that for the next five years they had belonged to mobile child units. One cautiously tells me that Ta Mok is now considered a criminal and Hun Sen a hero, but that in his opinion, there should be statues of all the leaders. His friends nod in agreement. All the leaders were fighters, they say.

**SEVERAL MONTHS BEFORE THE VERDICT,** I attended a Khmer Rouge victims' meeting in the south of the country, where I heard a number of odd suggestions about how best to commemorate Duch's crime. A young member of one of Cambodia's biggest human rights groups suggested that three statues of Duch should be raised at S-21: one of him seated in the horseshoe-shaped dock; one of him saying sorry; and the third of him being tortured. The well-intentioned young man went on to say that, on the day of the verdict, Duch should kneel before the ninety civil parties. Later, at the end of the trial, a Cambodian lawyer representing some of the victims also felt that a statue of Duch should be built at S-21, this time depicting him in his revolutionary uniform, that is to say, his criminal uniform, he wisely qualified.

Meanwhile in Anlong Veng, Nhem En also wants to build statues of Ta Mok, Pol Pot, and all the top Khmer Rouge leaders.

"We could inscribe them with 'good leader' or 'bad leader,'" he says.

"What would you write on Ta Mok's?" I ask. Nhem En and the four wedding guests laugh heartily.

"I can't give a definite answer," says one of them. "Most people say he was bad, but I'm not convinced."

"I would say 'bad,' because he lost the war," says Nhem En.

"And what about Pol Pot?"

One of the men tells me that he would put down "bad" for the same reason that Nhem En had given: Pol Pot lost the war. Then, as the conversation continues, everyone feels more comfortable and becomes more expansive.

"We should write 'bad' because millions of people perished and the country was destroyed," adds the same man.

Until this moment, he claimed not to know very much about the Khmer Rouge era; now, he gives a detailed list of the crimes committed under the regime. After this unexpected indictment, I again ask: "So, then, was Ta Mok good or bad?"

"I would say bad," he answers.

**ON THE STRAIGHT ROAD** leading from Anlong Veng to the Dangrek Mountains and the Thai border sits the Sra Chhouk pagoda and its impressive but unmarked mausoleum. It is the final resting place of Ta Mok. If the importance of former members of the Angkar can be measured by the size and upkeep of their tombs, then Ta Mok is hands-down the most powerful dead member of the Khmer Rouge, one of the few leaders not to have attended university. His grave stands as a sort of testament to his peasant superiority over the Revolution's professors.

Along one side of the property runs a long, incomplete wall made of concrete sections of equal size, each one with an inscription in Khmer. Each inscription, I'm told, indicates a donation. A pleasant, artificial pond ringed with eucalyptus demarcates the other side of the property. When I first visited in 2007, there was a platform of some six by ten meters rising about fifty centimeters off the ground, partly walled in by little columns made of faux mother-of-pearl and blanketed in flowers. A fine wooden roof covered this impeccably maintained kiosk. Rising from the middle of that square slab beneath it was a big, gray, rectangular cement tomb, about two meters across, with a slightly rounded top. No name or epitaph was engraved on it. The local authorities had permitted Ta Mok's house to be identified, but not his grave.

When I return three years later, the structure has been entirely redone. The simple wooden roof sheltering the tomb has been replaced with a real mausoleum of bricks and mortar. It looks a bit like an elephant, according to my motorcycle driver.

At a point where the path leading to the sepulcher intersects with another, one of Ta Mok's four daughters keeps a small shop. Inside, a handsome portrait of her father hangs on the wall.

"Whether he was a good man or a bad man, a father is a father," she says calmly.

She tells me that she spent the previous two years having the tomb rebuilt at a cost of $16,000. For the moment, she's too broke to finish the renovation, she says. But later there'll be a painting of the dead man, with his name beneath.

"Would you like to contribute?" she asks with a smile.


NEXT TO A LONG, narrow path nearby stands a humble wooden house on stilts. It's so small it's easy to miss. In front of the house, about a meter from the path, there's a low rise covered in pansies and shrubbery. There's no sign here, no guard, no one selling tickets. But since 2006, the provincial authorities have been protecting a plot of fifteen hundred square meters around this house, and have forbidden anyone from living there. Beneath that clump of earth lie the ashes of a dozen people: Son Sen, the dreaded minister of defense and founder of S-21, and his family, all of them murdered in 1997 at Pol Pot's orders, after he came to the conclusion that his faithful accomplice was, in fact, an agent of Hanoi. Son Sen and his family comprised the last "line" to be erased by Brother Number One. Of all the now-dead mass murderers who blighted the Cambodian people, Son Sen is by far the most neglected.

Leaving behind Ta Mok's opulent mausoleum and Son Sen's anonymous burial plot, the road leads straight toward the rocky ridge of the Dangreks, up an escarpment and to a small border post in the Choam-sa Ngam pass, fifteen minutes' drive from Anlong Veng. In the middle of an uneven rise, the road splits around what used to be an imposing statue, one that has become a site for offerings. Some say that the Khmer Rouge carved it to celebrate their victory over the Vietnamese, and it's been largely destroyed since. All that remain are a pair of legs and part of the torso of a revolutionary cadre. Tourists are encouraged to take photos of themselves resting their heads atop the decapitated bust. A dozen or so little shrines, called spirit houses, now surround the outcropping. The local population has transformed what was once a monument to the glory of the Revolution's soldiers into a sort of oracle, where women offer fruit and light incense. One of the amputated sculptures is draped in colorful fabrics. The locals have turned it into a god of fertility.

The atmosphere at Choam-sa Ngam is typical of border crossings, with its assorted traffic carrying everything from traditional market goods to the most corrupt forms of commerce.

There's a track running alongside and slightly below the main road. Next to the track, a short, narrow plank leads across the gutter to a field. Dirty water and detritus stagnate beneath this makeshift bridge. Head fifty meters down the sandy track beyond it and you'll see the sheet-metal shelter over Pol Pot's ashes. The house where the tyrant spent his final days, ill and watched over by Ta Mok's men, used to stand next to the path, but there's nothing left of it now. It was burned to the ground. Only a concrete slab remains, its cracks colonized by tall, cottony grass. Under a tuft of green, you can make out a fragment of an enamel toilet basin. I have a friend accompanying me, an American who specializes in museums and memorials for mass crimes. He amuses himself imagining this ceramic fragment becoming an artifact in an exhibition in New York or in the museum that Nhem En, who is following us down the path, dreams of opening.

"Pol Pot's pot," he says with a smile.

A gentle breeze adds to the serene atmosphere. I take a moment to enjoy my friend's Dadaist humor. Then, feeling discomfort and curiosity, I approach the despot's dismal grave.

A row of upturned bottles form a perimeter around the rectangular, ash-covered mound. A flimsy, broken wooden fence surrounds the small, rusted, corrugated iron shelter that stands a meter off the ground above the ashes, in which I can see a bit of tire, an empty bottle, and a brick. There's a packet of incense sticks and a little wooden elephant in front of the tomb. One thing is clear: there's very little upkeep. But the site has been duly registered by the government. A blue sign posted by the ministry of tourism tells visitors, in an English as broken as the French on the panels at S-21, to keep the place tidy:

PLEASE HELP TO PRESERVE THIS HISTORICAL SITE.
POL POT'S WAS CREMATED HERE. HELP TO MAINTAIN
PROPERLY. KEEP IT CLEAN.

That was in 2007. When I go back three years later, a row of red stakes has been planted along the trench. A brand-new sign marks the site. Nhem En has obviously pilfered the toilet fragment for his museum, or to sell to the highest bidder. Lovely pink and white flowers have bloomed in front of the mound, and a remarkable wooden spirit house, a sort of fine Khmer villa on stilts, protected on one side by a statue of an elephant, stands on a plinth. My Cambodian guides tell me that some Thais installed it after they won the lottery by betting on a number drawn from Pol Pot's biography. The appeal of legends lies in their perpetuation.

**ON A DIRT ROAD** in Choam-sa Ngam, lined with shops catering to tourists and travelers, there is a vacant lot strewn with wild grass and plastic bags. Fifteen or so gray wooden posts about three meters high rise from the lot. Now marked as a tourist site, it was here that Pol Pot's open-air trial took place before a small committee assembled by Ta Mok in 1997. From here, the road continues along a strip of forest on the Dangrek ridge line. In some places, it's no wider than a hundred meters between the cliff and the official border with Thailand. The path turns sandy, the sand white and fine. Not much farther on, we reach Uncle Roun's guesthouse. A rocky promontory directly overhanging the void gives a spectacular view of the great plain of Cambodia. This stunning belvedere seems like it was carved from the rock by man; the atmosphere is a little like that of some abandoned quarry. At sunrise, the lake of Anlong Veng gleams like a pool of molten metal. This was the Khmer Rouge's last central command post.

Lieutenant-colonel Roun joined the government's army in 1968, the year the guerrilla movement began its armed struggle and the year Duch was thrown in jail. Roun was sixteen years old at the time. He survived Pol Pot's four years in power by hiding his identity. When the Democratic Kampuchea regime fell, he rejoined the ranks of the government army. Following Ta Mok's arrest, he was sent here to guard the border. Dozens of soldiers, along with their families, settled on this ridge. In 2001, Ta Roun opened a guesthouse on this stunning spot. It has six rooms in which to host a few tourists as well as a number of regular locals who appreciate its location, a romantic hideaway perfect for carrying on affairs.

Out in the vegetation surrounding this pleasant guesthouse stands Ta Mok's radio antenna, as well as the house in which Pol Pot's men supposedly assassinated Son Sen. It's a small, one-room building. The walls are topped with wire mesh. My motorcycle driver tells me how, three years earlier, he had driven a Frenchman of about fifty here. Once they reached the spot, the foreigner took a can of black spray-paint out of his bag and sprayed enormous graffiti on top of the many messages that already covered the inside of the building. He wrote, among other things, the declarative SHAME ON TA MOK, the grandiloquent TA MOK, HISTORY'S MURDERER, and the more evocative TA MOK IS A COCKSUCKER. It seems that sometimes my countrymen travel far in order to express themselves.

The road that leads through the Dangreks passes through a lush forest that opens onto ponds, glades, and delicate marshes dotted with bare trees. Six kilometers on, the road shrinks and becomes first a steep path, then a track passable only by motorcycle. It rises a little, then widens a bit before reaching Khieu Samphan's "house." Samphan was at one time president of Democratic Kampuchea. Now, amid the trees and bushes, there remains nothing but the foundation slab, collapsed brick and concrete walls, and, on the ground, bits of green sheet metal that were the roof. The ministry of tourism has brazenly marked this crumbling and remote site with one of its blue signs. Below the site lies a large pond, on the far shore of which the path continues to rise. The land forms a sort of natural terrace here, protected on one side by the pond and on the other by a steep cliff. At its highest point, right at the edge of the cliff, lie the ruins of Pol Pot's final home, where he lived before his arrest.

The house was built in 1993. My guide tells me that when he first came here in 2005, the main floor was still covered with fine tiles. Most of them have since disappeared or disintegrated. There are two openings tall enough to pass through in the concrete wall. With a few acrobatics, you can slip through these to reach the basement. Here, you'll find two dingy rooms, both ten by three meters. Most of the house above has been demolished, with the exception of a raised terrace surrounded by a steel barrier and offering a superb view over the plains below. Several imposing cisterns of reinforced concrete have survived the destruction. A short distance beyond the house, a large water reservoir had been built so that water overflowed the precipice to prevent flooding. Three nonchalant young soldiers supposedly guarding the site are busy picking fruit from the trees. The ministry of tourism sign explains that it wants to "show Cambodia in all its glory."


**NHEM EN AND THE JEWELER** turned out to be right. Two years after our meeting, the road from Siem Reap to Anlong Veng was redone and is now one of the best in the country. It takes just an hour and a half to cover its 120 sinuous but perfectly paved kilometers. While the jeweler has pulled out of the museum project, Nhem En has stuck with it. Ten kilometers outside of town, where the road to Anlong Veng branches off the highway, stands a big sign indicating his museum some three kilometers away. Two portraits of Nhem En occupy the top left corner of the sign: one in his khaki Khmer Rouge uniform, and one in a suit and tie, which gives him the look of a local politician.

Tenacious and obstinate, Nhem En has clearly worked out how to make his government position pay. He's no longer deputy governor but instead a "district inspector." He now owns a vast stretch of land, some fifty hectares of rice paddies and fields subject to the Cambodian climate's vagaries. Under a ferocious sun, I find him busy planting multicolored flags at the entrance to his museum. He has a *krama* tied around his waist and a straw hat pulled down tight on his head. He tells me that the museum will be finished within two months. In the middle of a field baking in the blazing heat, his workers, both men and women, all of their bodies completely covered against the sun and all of them wearing *krama*s over their hats, have laid out the foundation of a building measuring twenty-five meters by seven. Nhem En hopes to draw a thousand visitors a month.

"How much will the entrance ticket cost?"

"I don't know."

Nhem En has had two magnificent *baray* (irrigation reservoirs) built nearby. Overflowing with enthusiasm, he insists on taking me and my two friends to see them in his car. Though they're less than a hundred meters away, we drive around them. A tractor quickly flattens a track ahead of us while we follow. It's the sort of comical, absurd scene that wouldn't be out of place in a Charlie Chaplin film. With his high-pitched voice and machine-gun delivery, Nhem En tells us about the restaurant he envisages building alongside the guesthouse, where the bungalows will have names such as Pol Pot, Nuon Chea, and Khieu Samphan.

"Tourists will be able to choose which one they sleep in," he says with a big smile and sunlight glinting in his eyes. "The first thing people want to see is Angkor Wat. The second thing is the Khmer Rouge."

Nhem En thinks he can do as well as the temples. He takes me by the arm and convinces me that, at the very least, he isn't short of ideas. "We could exhume Ta Mok's body and put it on display in the museum. People would pay $20 to see him. He would be behind a glass partition that would open electrically. We could raise him with a hoist. We could also display his wooden leg, if I can find it."

Nhem En is beaming. He reaches out, strokes my belly, and affectionately takes hold of my chin. He is absolutely delighted by our visit.

# CHAPTER 26

**P**EOPLE ALREADY VISIT S-21 AND CHOEUNG EK. MORE AND more will visit Pol Pot's tomb. But no one will ever visit Prey Sar, known as S-24 under the Khmer Rouge and an integral part of the concentration complex under Duch's authority. Prey Sar was a prison before and during the Khmer Rouge years. Prey Sar is still a prison today. During the revolutionary period, it was called a "reeducation camp."

"The long-term purpose of S-24 was to smash prisoners, so the term 'reeducation' was just political-speak, wasn't it?" a judge asks Duch.

"That was the general idea. Everyone could see it. The Revolution's aim was to smash them progressively, one by one."

In many ways, the entire nation of Democratic Kampuchea was turned into one vast forced-labor camp. S-24 wasn't so different from the many cooperatives established throughout the country where Cambodians died from starvation, illness, or exhaustion by the thousands. What distinguished S-24 from the other labor camps was that its inmates were people who had committed "infractions." If other cooperatives were run on the basis of class, most prisoners at S-24 were Communist Party combatants. Yet again, the tribunal metes out justice on behalf of those who, for the most part, had served the regime before falling victim to its purges, rather than on behalf of the masses of ordinary people whose lives it destroyed.

S-24 was where the Party parked those combatants it deemed ill-disciplined enough to lock up but not enough to execute—at least, not right away. It was for those whom the Party hadn't yet decided whether to wipe out or not. The prisoners at S-24 were referred to as "elements." Their job was to grow rice and cassava and raise animals. There were around thirteen hundred elements at S-24, divided into three groups. The least troublesome were sometimes released and re-integrated into their combat units if they behaved well, worked hard, and survived the inhuman conditions. The intermediate elements were held for evaluation. The serious elements were either worked to death or exterminated at Choeung Ek. According to the surviving archives, 590 S-24 prisoners, including 50 members of its own staff, were transferred to S-21 and killed. We don't know how many S-24 prisoners died in the camp or were sent directly to the execution fields.

Duch says he can remember having gone to S-24 four times. The last time was in December 1978, a few weeks before he fled Phnom Penh. He had gone there to arrest his deputy, Nun Huy, the head of S-24 and one of S-21's three-man leadership team. Duch says he had many reasons for arresting Nun Huy, but has forgotten most of them. The main reason was that a radio operator had run away, he says. "I asked my superiors that he be arrested. Nuon Chea agreed to it."

Nun Huy was planting potatoes when he saw Duch coming for him. His wife, Comrade Khoeun, the deputy head female interrogator at S-21, was arrested the next day and executed. Yet another "line" erased.


**IN 1977, BOU THON'S HUSBAND,** a motorcycle chauffeur named Phuok Hon, was arrested and never seen again. Though he had been a low-ranking soldier in the Revolutionary Army, Phuok Hon had had the misfortune of having been introduced to the Revolution by a high-ranking Party cadre called Koy Thuon, who was both Cambodia's minister of commerce and responsible for its northern region. Koy Thuon fell from grace in 1976. By the beginning of 1977, he found himself in S-21. Phuok Hon fell in the wake of his mentor's fall, one more dot on Koy Thuon's line; one more domino in that bloody game of traitor networks that Duch was in charge of unearthing and eliminating. Three months after her husband's disappearance, black-clad guards came to tell Bou Thon that her husband had stolen some petrol; they told her she could go to him. Instead of being reunited with her husband, she was put in prison and beaten. Pol Pot's agents decreed that when a soldier was eliminated, his wife and children were, too. While Phuok Hon was sent to S-21, where his photo remains to this day, Bou Thon was sent to S-24.

Bou Thon and Phuok Hon had four children, all of whom died. In court, she tries to overcome her pain while villagers stream into the public gallery. There are some new faces on the bench reserved for civil parties: the parents of victims of S-21, who have come from abroad to give statements the following week. From their bench, Duch appears so remote that his head, visible over the top of the lectern, looks like a durian fruit sitting on a shelf. A security guard slaps a mosquito as Bou Thon begins describing everyday life for convicts during the Revolution. At first, her grim litany of Cambodia's collective enslavement wears me down and I find it hard to focus. The trial has been going on for four and a half months now. You get inured to it. Some mornings, you even feel oddly cheerful.

But Bou Thon speaks with great intensity, as though she's longing to speak. At sixty-four, with beautiful gray hair, she has grown out of any shyness. She wears a white blouse and a silk gray-green-and-gold scarf. Her sophisticated elegance makes it easy to forget that she's illiterate. She smiles. She, too, seems oddly cheerful.

In contrast, Duch's wrinkles and the bags under his eyes appear especially pronounced this morning; he looks tired after so many difficult days, like one of Marcel Marceau's sad mimes. But he doesn't hesitate to corroborate Bou Thon's identity, which he does in his most honeyed voice. He has no choice: her biography and photograph are in the prison's archives.

"My husband used to say that he didn't want to live under Pol Pot because there was never enough to eat," she says.

Bou Thon was living in Phnom Penh in 1973. When Pol Pot's men seized power, she found she was the wrong type of person, a city-dweller. In court, Bou Thon talks and talks and talks. But instead of making me drowsy, her story grips my attention. She uses fragments of dialogue and description to recreate the banality of camp life, and to evoke how the twin obsessions of hunger and interrogation hung over everything. Her anecdotes are precise and detailed, and I sometimes wonder why her memory favors some stories over others. Why she remembers the banana story, however, is obvious.

One day, while working in the fields, Bou Thon saw a cluster of bananas. She said out loud that they would be good to include in the daily meal. That was enough for the guards to accuse her of being the enemy. Her face still bears traces of the beating they gave her.

Every day, the "elements" planted rice, carried water, checked the corn, and cultivated vegetables. Their work quota was strictly monitored. All production went to the mysterious, shadowy Angkar. "We weren't allowed to eat what we grew. And we didn't dare protest. They had complete control over me. We had to obey their orders. We weren't allowed to question them. They could've decided to kill me whenever they wanted to."

No one spoke to anyone else. Everyone had watched someone get dragged away, never to be seen again. In this prison without walls, no one said a word.

Until the day Vietnamese troops invaded Phnom Penh.

"We ran for two days. I was loyal. I stayed with the group. I was stupid to follow the prisoners. I followed blindly. I don't know why. We reached the town of Amleang. We found shelter there for two nights. People said that was him, the prison director. Of course I knew him!" she shouts, sitting tall in her chair. "He was a small-framed man"— Duch smiles—"and I even knew his wife. She was tall and well-built."

Shortly after her escape, Bou Thon found out that all her children were dead. She went back to her native village. Going home frightened her. Being alone frightened her.

"I suffered so much," she says in a hushed voice. Her voice breaks beneath a tide of tears.

*My uncle is the head monk in our village. He advised me to try*
*and forgive and forget. But when I was working in the fields and*
*paddies, I would ask myself: "What's the point? There's no one left to*
*work for." My mother also told me to try to relax. But I want to be*
*here, at this tribunal. I want to be here so that justice is done on behalf*
*of my husband and my children. Why were my children executed?*

The court declares a recess. Bou Thon pours herself some water.
Duch lingers behind his table, his mouth hanging slightly open, wearing that mask he puts on when he's troubled but doesn't want others
to know it. He watches the people leaving the public gallery. Then he
turns to his Cambodian lawyer and assistant, both of them relaxed
and smiling, and he seems to immediately recover his work attitude,
standing with his hands in his pockets.

Bou Thon was no Khmer Rouge cadre. Prey Sar is the forgotten
place, the relegated place, the place from which, on July 23, 1977, 160
children were sent directly to Choeung Ek and executed because the
Party feared that they would grow up and seek vengeance, and because they were too expensive to feed. Prey Sar hasn't been turned
into a museum like S-21. Today, Prey Sar is still a prison, just like it
was before Pol Pot, when Duch, Mam Nai, and Pon were incarcerated
there. In court, Bou Thon's testimony focuses everyone's attention on
Prey Sar, and for a day the prison is remembered. And then it is forgotten again, consigned to history's oblivion.

Duch starts to speak and Bou Thon begins to weep. Her sobbing
triggers Duch, who starts crying, too, shedding long-repressed tears.

"The tears that fall from my eyes are the tears of innocents," sobs
the former director of S-21 and S-24.

*I want to be close to the Cambodian people; if they choose to*
*condemn me, they can and I will accept it. I must accept it, no matter*
*how heavy the sentence. I won't use a bucket to hide an elephant. At the*
*time, we thought that the Yuon [Viets] were invading Cambodia. Now,*

*finally, before all of you and before the Cambodian people, I would like to share this pain from the bottom of my heart. I will accept this court's judgment. I wish for the Cambodian people to condemn me as quickly as possible.*

Total silence fills the public gallery. A rustle, the sound of people brushing against one another and the sounds of joints cracking betray the first hesitant movements of people in the gallery. Duch had them rapt, stone-still, and now they're surprised and a little embarrassed for it. Duch salutes the court, then bows low to the public gallery. With her carefully groomed hair, elegant sarong, and dignified bearing, Bou Thon walks quietly from the room.

# CHAPTER 27

**W**ITH THE VIETNAMESE APPROACHING, PEOPLE STAMPEDED out of Phnom Penh. Duch didn't spare a thought to the fates of the dozen prisoners—the painters and sculptors, the electrician and mechanic and dentist—whose temporary usefulness to the Party had delayed their execution.

> *When Nuon Chea ordered me to empty the place, I didn't think about those people I was keeping for my own use. I never imagined that the Communist Party of Kampuchea would be overthrown. When the Vietnamese got close, I fled, leaving the prisoners behind. That's why they survived. Not because I had pity on them or had a plan for them. I simply didn't think about them.*

"Eventually, they, too, would've been smashed, correct?" asks a judge.

"That's correct, Your Honor."

Duch left S-21 on foot, late on the morning of January 7, 1979, with Vietnamese tanks on his heels. Like Bou Thon, he walked for two days without food or water. The fates of the dozen prisoners who worked at S-21, or those of the "elements" at S-24, or of its staff, no longer concerned him. From now on, it was each man for himself. Duch fled northeast, toward Amleang. They say a murderer always returns to the scene of his crime. For a while, Duch took refuge on the former site of M-13, the place where his career in the revolutionary police had begun. The men who had then been under his command were now under that of Ta Mok or had scattered or been killed.

Duch ended up in Samlaut, in the northwest of the country, under the command of the military chief Sou Met with whom he had worked before, during the purges of 1977. There are nine damning letters in the S-21 archives, in which Duch sends the Khmer Rouge division commander information acquired from prisoners' confessions. Around three hundred of Sou Met's men were eliminated at S-21. Sou Met later achieved the rank of general in the national army. He is now retired peacefully in the quiet town of Battambang. The government opposes putting Sou Met on trial. No one involved with Duch's trial—not the prosecution, defense, civil parties, or judges—has seen fit to summon him. Justice, in this court as much as in any other, bends to political power.[*]

Duch describes how he became head of transport for the region, how he refused the offer to command a division, and how he became a sort of private tutor to Sou Met's children. It was at this point that his career as a torturer ended as suddenly as it had begun less than eight years earlier. Duch was still a member of the Khmer Rouge, but he no longer held any position of responsibility within its security apparatus. The man once described as one of the most powerful, enthusiastic, and well-connected officers of the Khmer Rouge, privy to the Politburo's most sensitive secrets, became a nobody. He sidelined himself. Once again it was the guerrillas and their war that mattered, not the police and their purges. Duch avoided taking on a military role. No one in the courtroom has properly explained why his political career ended so suddenly: neither the prosecutor, nor Duch himself.

## "I LIKE TO HUNT BIRDS."

In 1980 or 1981 (he's no longer sure), Duch was in his secluded house in Samlaut, cleaning the rifle he used for hunting, his new pas-

---

[*]   Sou Met's death was announced on June 26, 2013, twelve days after he had passed away, at the age of seventy-six. He had been under confidential investigation since late 2009, and was never charged.

time. He assembled his weapon, put a bullet in the chamber, poured a little lubricant in the magazine, and pointed the barrel toward the sky. His wife, wondering how he could assemble a weapon in such manner, suddenly reached out and put her finger on the trigger. The gun fired and Duch's hand was partially torn off. His wife, a nurse, quickly wrapped it up in a bandage and took him to the nearest clinic. He had to have a finger amputated and, ever since, his mutilated left hand looks sort of like a flipper when he waves it about in that curt, martial way of his. Once, during the trial, a journalist friend whispered to me that she thought it looked like a bird's foot. Duch stopped hunting birds after the accident.

In the mid-1980s, Duch went back to teaching. He taught at a primary school in what was then still Khmer Rouge territory. His wife gave birth to two more sons, in addition to the two children born during Duch's years at S-21. Son Sen asked him to go to China and teach Khmer literature there. Kaing Guek Eav, a.k.a. Duch, asked to change his name again. He became Hang Pin. The name *Hang* is from his Chinese clan, and was the perfect introduction with which to go and teach in the land of his ancestors. For Duch, *pin* meant "lazy student," the opposite of *duch*, "good student." The Buddhist Institute Dictionary, on the other hand, defines *pin* as "summit" or "superiority." Thus, *pin* connotes both the useless person Duch became and the superior one he always aspired to be.

He took off for Beijing in September 1986. When he returned two years later, he worked under the supervision of Son Sen's wife before taking charge of the economic affairs of the village of Phkoam. Peace talks were taking place at the time, and the UN was overseeing the establishment of multiparty elections in the country. Duch was there to prepare the campaign for the Khmer Rouge faction. In the end, the Khmer Rouge rejected the 1991 peace accord and resumed armed combat. But its fighters only lost ground and Phkoam fell under government control. Duch (going by Hang Pin) lost contact with Party headquarters. He no longer believed that the Party would win. Like many Khmer Rouge soldiers, Duch started thinking about ways to get out, though he was careful to maintain his relationships with his peers. The Revolution was dying. The Brothers were starting to claim that the Vietnamese had invented S-21. Son Sen's wife said the Swedish experts who examined the prison grounds established that the corpses found there belonged to people who had died after January 1979. This was one lie too many for Duch. In the universe of untruths he had inhabited for twenty-five years, he had reached his limit. Worse, he found himself being stripped of his biggest contribution to the Revolution. "I couldn't really accept it, because they were talking about history. I knew the real history. How could anyone say that S-21 was a Vietnamese invention?"

Duch returned to the institution where he got his start before he had set out to become a New Man: the public education system. He taught physics and chemistry at the high school in Phkoam. Then, in November 1995, under circumstances that aren't entirely clear, Duch's wife was stabbed to death during a burglary at their home. That part of the country was still dangerous: there was combat nearby, and crime was endemic.

"There was a great deal of instability at the time. Every day, there was gunfire and robberies. No one obeyed the law," says one witness.

Duch now believes that he had been targeted by his dreaded enemy, Ta Mok. He describes his wife's murder without a trace of emotion, just as coldly as he described his brother-in-law's murder within the prison walls. When he speaks of his mother, who's still alive, or his father, who died in 1990, he also does so in the most detached tone imaginable.

"I have never dreamed of my father," he says. Nor does not seeing his children appear to provoke any sign of sadness in him.

"Disempathy is the inability to think another person's thoughts and to feel another person's emotions; it's the inability to imagine that others are different from one's own self," explains the psychologist. "There are signs of it in Duch, who killed off his own personal identity in order to identify with a collective one. Yesterday, it was Communism; now, it's Christianity. This disempathy is neither absolute nor total. Another of his characteristics is what psychologists call alexithymia. This is a clinical concept that designates, in this case, the defendant's inability to both consciously feel emotions and articulate them. This can't be blamed wholly on Sino-Cambodian culture, though there are contributing cultural factors. Duch thinks pragmatically. He thinks about what is 'practicable,' to use the word he employed."

**CHRISTOPHER LAPEL IS A PASTOR** based in Los Angeles, California. He was thirty-seven years old when, one day in late December 1995, he met Hang Pin during one of those evangelical "crusades" that the West has been sending out into the Third World for the last thirty years with an energy not unlike that of the swarms of locusts that fall upon the crops of the Sahel. The pastor knew nothing about the past of the man he was welcoming. But what did it matter? Hang Pin, suddenly and violently a widower, wanted to give his life to the Lord. He came across as a kindhearted man, a servant of God, warm and welcoming. On January 6, 1996, after two weeks of religious instruction and seventeen years to the day after having shut down S-21, Kaing Guek Eav, alias Duch, alias Hang Pin, was baptized in the river. He was declared ready to preach. Pastor LaPel immediately instructed him to go out and spread the Good News. Three years later, when he learned of his new convert's real identity, the pastor was surprised but happy. "It fills me with joy to see God transform a man's life by turning a killer into a believer."

According to Communism, loving others means giving the proletarian class an absolute monopoly on power, says Duch. When he lost his faith in Communism, he realized that Christianity's teaching to love one's neighbor as oneself could be beneficial. "I thought converting to that religion would be a good idea because it allows you to love your enemies. I wanted to make sure to only follow a religion that treats everyone as equals, with no regard to class. I believe that Christianity is that ideal religion."

Confession in Communism, confession in Christianity: in both cases, salvation depends on a preliminary interrogation during which, Duch knows as well as anyone, one's ability to conceal is paramount. Does baptism, which washes away sin, presuppose total sincerity? Fortunately for his mission, Pastor LaPel doesn't appear too concerned about such scruples. Duch appeared at the mission of his own volition, saying he understood the meaning of baptism and was ready to receive Jesus Christ as his Lord and Savior. He was baptized and saved. Amen.

> I remember when I first met Duch, or Hang Pin, in late 1995. I saw a man steeped in sadness, living without peace or joy or any purpose in life. But I saw him change completely after I baptized him. He turned around 180 degrees and is a completely different person now. He's a man who lives in peace, who lives with joy and purpose. I remember his baptism very clearly. He looks like a different person now. He dresses well and wears glasses. He listens, preaches, and teaches; he asks questions about God, Jesus, and the Holy Spirit; he wants to know about sin and redemption . . .

Hang Pin returned to the village to open a new church for fourteen families. From *homo sovieticus* to born-again Christian, his miraculous journey continued. It turns out that Judge Lavergne, however, is not a believer. He's confounded by the speed with which one can be ordained a lay pastor. "It seems to me that Duch's conversion happened extremely fast. Do all the people you baptize convert as quickly as that, Pastor LaPel? Or was Duch an exceptional case?"

> It's hard to describe. When they hear the Word of God, it pierces their hearts. I have no control over what happens inside their hearts after that. After Duch received the training and learned the Word of God, he couldn't wait. I could see how impatient he was. He wanted to go back to his village to bring Jesus Christ to his friends and family.

"So his mission was to share his newfound faith in Christ with his friends and neighbors. Did his mission go beyond that? Could he teach? Could he baptize people? Could he hold services?"

"Yes, Your Honor. He had the opportunity to lead worship services. He had the opportunity to teach the Word of God. As leader of his church, he had the opportunity to pray and to receive communion with the believers in his village."

"Were you never concerned that there might be shortcomings in this recent convert's faith?"

"I'm not concerned insofar that, so long as he teaches the Word of God, people experience the Word. That is the only answer I can give you, Your Honor."

"Can you tell us how many people have been baptized into your church here in Cambodia? How many people belong to it?"

"I have lost count. There have been many thousands of new believers during my trips to Cambodia over the past eighteen or nineteen years."

"Thank you, Pastor LaPel. I have no further questions for the witness, Mr. President."

The prosecutor asks:

> Did you not ask yourself whether Duch converted for
> psychological comfort, or because it was convenient, or for pragmatic
> reasons? Did you not wonder, for example, whether he converted
> in order to benefit, immediately and unconditionally, from the
> forgiveness offered by the Christian god, and so avoid having to face
> the Buddhist cycle of reincarnation?

"It is difficult to grasp the power of Our Lord Jesus Christ," replies the pastor, perhaps sensing the general disbelief in the court.

The prosecutor gives up. Another judge takes over.

"Were you not concerned that in this entire process you had been manipulated from the outset by someone who had hidden his real identity and his past from you and who, furthermore, is responsible for the deaths of more than twelve thousand people at S-21, including people very close to you?"

> *I lost close friends at S-21, and I lost my parents, my brother, and my sister in the killing fields. When I saw Duch again in June 2008, I told him that I loved him and that I forgave him for what he did to my parents, my brothers, my sister, and my close friends at S-21. I spoke for myself as a Christian, as a follower of Jesus Christ. In that moment, I felt peace. I was filled with joy. When I told him that I forgave him and that I loved him, the healing was mine. I hate the sin, not the sinner. When you are a true believer, when you understand the Word of God, you understand what forgiveness truly means. I speak simply as a believer.*

There are at least two reasons why Maître Roux, who hails from the predominantly Protestant Cevennes region in France, might understand Pastor LaPel better than others do: his Protestant background and his duty to defend his client.

"Would you agree with the notion that, above and beyond any theological training that a person might acquire, what matters most is a person's direct, intimate contact with God, and that such an encounter doesn't necessarily require in-depth study?"

The pastor agrees.

"Does the Bible not contain examples of people who converted very suddenly, some of whom had committed crimes?"

The pastor can't seem to find any. He seems content to simply reiterate that "when a person gives himself to God," etc. The man of law tries to guide the man of faith.

"To be precise, isn't it true that every man and woman might one day travel his or her own way to Damascus?"

But Paul of Tarsus doesn't seem to strike a chord with, or matter to, the pastor. Christopher LaPel's theology, like Communism's, doesn't require any great learning. It requires only that a person understand what happens when he receives and understands the Word of God (or the Party), and that a person give thanks and praise. All that counts is absolute faith.

The pastor is proud of his former parishioner for admitting to his crimes and presumably accepting his punishment. He encourages Duch to continue spreading the Word of God, because "Jesus is the Prince of Peace and our only answer." Duch stands when the pastor leaves the courtroom, a gesture of respect toward his new master.

**ONE OF THE TRIALS** I followed while covering the international tribunal for the 1994 Rwandan genocide was that of a kind young man, a civil servant with the Belgian social security service, who went to Rwanda and turned into a murderous, rabble-rousing radio presenter who supported the militia in its quest to systematically exterminate every Tutsi it could find. After the *génocidaire* forces were defeated, the international authorities went looking for this Belgian, who changed not only his identity but his religion. Originally Christian, he converted to Islam. The ideology that had caused him to commit the crime in the first place was discredited. Now it gave way to another cause, which the man embraced with equal passion. Just as this Belgian has forsaken his faith in Hutu Power for that of the teachings of the Prophet, so Duch renounced Communism for Christianity. Many people experience an inescapable urge to believe. Fortunately, most don't find themselves in the kind of extreme historical circumstances that might lead them to follow that urge to its worst possible end.

Duch told the psychologists, "You can't live without faith. At first I believed that the Communists would save my homeland. Now I know that it's God."

He says that religion is what defeated Communism in Poland and elsewhere behind the Iron Curtain. He also hoped that by converting, he would escape his karma. Furthermore, it meant that he would be joining the winning team, just as he does by cooperating with the international justice system. Yet the psychological expert sees another side to Duch's conversion, one that bears witness to something more profound than an expedient shift of allegiance.

> *He changed group. The choice of group is certainly an interesting aspect of the defendant's development, but what is most interesting is that he chose a group that allows for the individual. In Christianity, each person has direct contact with God. So it's a completely different system. In Communism, the individual disappears. But now he has joined a group where the subject exists in his or her union with God. It seems to us that joining this group is of some therapeutic value to the defendant.*

From 1996 on, Hang Pin the Christian resolutely endeavored to be as much like the pre-Revolution Kaing Guek Eav as possible. His boss at the school speaks of him in the same glowing terms as his former students and his classmates from the 1960s did: Duch is a humble, meticulous, and hardworking man, much appreciated by his colleagues. He is punctual, rigorous about deadlines, has fine teaching skills, and never talks politics. Duch—biddable, discreet, irreproachable—melts into the community. "He was gentle and he didn't talk much. He did what was asked of him; if someone asked him to clean, he cleaned."

At the end of 1996, Duch was asked to teach French and was entrusted with distributing textbooks. He was an ordinary man again. He had aged. At fifty-four, he was an elder, and people called him *krou ta*—"master" or, literally, "grandfather teacher." Duch smiles while the civil servant tells his story.

Whenever a former colleague of Duch's or his former head teacher is in the courtroom, Duch stands respectfully, watches intently, nods in agreement. He always has a smile for witnesses who knew him before or after he was Duch. None of them are tormented like those who suffered under Duch. Their perception of Kaing Guek Eav neither frightens nor intimidates them.

Then one day, without warning or explanation, Hang Pin disappeared. The school's head teacher learned that he had gone to work in Samlaut.

"Even today, I find it hard to accept that he was implicated in these crimes. He was such a perfect teacher. He was gentle, generous, and friendly. He was exactly the opposite," the man says.

The defection rate among the top-level Khmer Rouge leadership accelerated. In 1996, Brother Number Three, Pol Pot's brother-in-law, deserted, taking with him thousands of men. In June 1997, Pol Pot killed Son Sen, Duch's mentor. In Samlaut, Duch hoped that his superior, Sou Met, would take him along when he surrendered with his men. But Duch was stuck with the last remaining rebels of Democratic Kampuchea. Pol Pot died in April 1998. In August, Duch was finally repatriated to government-controlled territory. Toward the end of 1998, Brother Number Two, Nuon Chea, Duch's boss at S-21 from 1977 to 1979, surrendered. Thirty years after it had started, the civil war was over. In March 1999, the Khmer Rouge hardliner Ta Mok was arrested and jailed. Hang Pin was expected to take charge of education at a district level. He readied himself for his new promotion. Around that time, Kaing Guek Eav, alias Yim Cheav, alias Duch, alias Hang Pin, became a grandfather and seemed at least partially reconciled with his identity at last. He gave his real name, Kaing, to his grandson, as well as his Chinese name, Yun: Kaing Yun Cheav.

"But everything was compromised when Nic Dunlop found me," he says, struggling to hide his bitter regret.

Nic Dunlop is a charming, thoughtful, and reasonable man. So much so it is funny to imagine this young Irish photographer's curious habit of carrying what was then one of the few existing photos of Duch around in his pocket during his trips through the region. One day in April 1999, on the outskirts of Samlaut, the tenacious journalist recognized the former head of S-21. The chameleonic Hang Pin was working for an American refugee-assistance organization. When

Dunlop confronted him with his past, Duch immediately conceded the truth.

"I firmly believe that nothing can be kept secret forever. You can keep something secret for a while, but not for very long," he tells the court.

On May 10, 1999, Duch was arrested and transferred to the military prison in Phnom Penh.

# CHAPTER 28

**W**ERE CIRCUMSTANCES DIFFERENT, MAM NAI WOULD BE SIT-ting next to Duch in the dock. He was part of the entire murderous undertaking, from M-13 to S-21. Though ten years older than Duch, Mam Nai remained his subordinate throughout the Revolution. Mam Nai isn't and has never been a talkative man. Unlike many other former S-21 staff, he has steadfastly refused to talk to those researching the Khmer Rouge killing machine. Unapologetically severe, Mam Nai lives in the countryside. He grows surly the moment anyone criticizes the Revolution; he is not targeted by the purely symbolic justice required by the Cambodian government and opportunistically promoted by international law activists. At seventy-six, the only still-living chief interrogator from S-21 can spend his last years at ease, discreetly, and leave it to his former boss to deal with the burdens of shame and public remorse. Mam Nai has just one difficult moment to get through: he's been summoned to testify before the tribunal.

Mam Nai is about 1.75 meters tall, a lofty figure among his countrymen. On the first day of his deposition, he's wearing a dull, bottle-green jacket. The Khmer Rouge used to wear *krama*s thrown casually around their necks, the ends hanging down their chests. The *krama*s were the only colored items of clothing allowed with the otherwise strictly black revolutionary uniforms. In court, Mam Nai always wears a *krama*.

Mam Nai's *krama* perfectly matches his jacket. He has suffered from a skin condition since childhood, and multicolored fingerless woolen gloves protect his hands. The hair on the top of his head and temples has grayed, while that on the sides and the back of his neck is still dark. Oddly, it looks as though someone has tonsured a circle the size of a tennis ball into the back of his head. The following day, he wears a high-necked blue-gray shirt, again with a matching *krama*. Mam Nai exudes an outdated, almost psychedelic, sense of style. In a rasping voice, his head and hands and upper body continually in motion with those slight movements common among the elderly, Mam Nai quickly makes it clear that he has no intention of divulging Party secrets. "I was assigned to interrogate low-value prisoners. That's all."

What about torture? "I don't know. From what I saw—and speaking in general terms, torture may or may not have been used—I didn't see any sign of it. I wasn't paying close attention. It's possible that torture took place. But I can't give any details about it."

Mam Nai's notebook was found at S-21. It is 396 pages long. It contains an account of staff meetings and training sessions. There are numerous references to the use of torture. This is typical of Mam Nai. He claims that no one ever gave him an order to torture a prisoner, that he confined himself to simply making notes of the meetings, and that he was and remains completely unaware of the other interrogators' methods.

What if the prisoner didn't confess? "I would tell the guards to take him back to his cell, to give him more time to think it over."

And what about the executions? "That subject isn't clear to me. I have no knowledge of this. The prisoners were simply taken back to their cells."

Mam Nai says he knows nothing about how S-21 was organized, nothing about the identities of the other interrogators or even how many there were, nothing about the conditions in which the prisoners were held, nothing about the number of them being held, including the Vietnamese, of whom he was in charge. "There might've been ten or twenty of them," he says.

What about Duch's instructions? "I don't remember anything about them."

*Ot dang te*—I don't know—he says curtly, over and over.

"We acted as though we could see or hear nothing, like the kapok tree," he says, referring to a tropical tree whose seedpods are covered in a silken, cotton-like fiber that is both water- and rot-proof. "That is why I am still alive today."

The public gallery reacts angrily to Mam Nai's denials. But the glass wall dividing the courtroom from the gallery prevents the witness from hearing the disapproving murmur that then rises into a collective nervous laugh.

"Do you have trouble with your memory?" asks an irritated Judge Lavergne.

"It's been a habit ever since my studies. I don't remember dates. Sometimes I can't even remember the names of my children," replies the witness, unabashed.

Mam Nai, aware of all the traps that interrogators lay, tries to keep his answers short. But attack him and he bursts to life, revealing a man still committed to the very beliefs that have misled him. "The country was under attack from American imperialists, therefore detaining prisoners was a necessary measure. Living conditions were horrible for all of us. Everyone ate the same thing, more or less."

Sometimes he blames the hierarchy: "As a subordinate, I was told to put it in writing, that's all. Of course that's my writing. But I didn't know what tricks my superiors were up to. I was their subordinate and I simply had to follow orders."

Or else it's on account of that vexing problem, a poor memory: "Do you remember anything about these documents, Mr. Mam Nai?" asks the judge, referring to interrogation reports bearing Mam Nai's signature.

"I can't remember. 'Chan' was my name, of course. Yet as hard as I try, I can't remember."

**THERE IS A WELL-KNOWN PHOTO** taken during the S-21 years: four men, including Duch and Mam Nai, stand behind their respective wives and children. Obviously, Mam Nai is obliged to admit to the court that he recognizes his former boss in the photo. But he has the nerve to say that he doesn't recognize the other two men. When the court shows him documents proving that he had interrogated a Western prisoner, Mam Nai asserts his right to silence. Mam Nai isn't on trial, and never will be, so he can afford to say nothing. He makes the most of the blessing of amnesia. All he has to do is get through an unpleasant forty-eight hours, the only time in his life when he will be the object of public scorn. He rubs a little Tiger Balm into his neck and behind his ears, cleans his spectacles with his *krama*, and weathers the storm.

Someone asks if he has any regrets. "My regret is that the country was invaded. First the United States invaded us, then the Vietnamese. That's what I regret."

When court officials were reconstructing events at S-21 during the trial's investigative phase, Mam Nai would show up wearing a funny-looking beanie. It had three stripes: one purple, one red, and one orange, and the strangely provocative words—NO FEAR—written across its front. He claimed not to know what they meant.

Nothing is more chilling than a Communist intellectual. Because an intellectual is, by definition, a member of the bourgeoisie, he must persuade others of his proletarian transformation.

"Every Party member who was not from the peasant class and with no connection to someone powerful had to work relentlessly to meet the eligibility criteria," explains Duch.

The intellectual who feels obliged to apologize for his pedigree compensates by adopting even more radical ideas. Mam Nai believes his own metamorphosis was successful. "I was a teacher, I was from the bourgeoisie, but I adapted myself to the proletariat. I succeeded. That's why the Party accepted me as a member."

"I have no further questions for the accused . . . I mean, for the witness, Mr. President," says Judge Lavergne, making a slip that gives some relief to the bitter indignation simmering deep in the hearts of everyone in the public gallery.

**FROM THE END OF 1977,** the armed conflict between the Khmer and Vietnamese Communists escalated, and many more Vietnamese prisoners started arriving at S-21. According to the available archives, 345 Vietnamese soldiers, spies, and ordinary citizens were interrogated and then eliminated at S-21. In the propaganda it broadcast via its radio station, the Khmer Rouge called on Cambodians to kill every Vietnamese they found. In 1978, the Khmer Rouge made films of those Vietnamese it had captured and sent to S-21.

"If we wanted them to say that in Vietnam people were dying of hunger, then they said it. Their confessions matched what we wanted to hear," says Duch.

Duch struggles to hide the visceral dislike he and a considerable number of his countrymen harbor toward the *Yuon,* a pejorative term for the Vietnamese. He makes a point of telling the court that he doesn't use that word derisively, but he stiffens when a specialist of the region describes the two age-old neighbors as "brother enemies."

"Did you hate the Vietnamese troops?" asks a judge.

"That was my feeling at the time. It's an old story, a long-standing quarrel."

Whenever Mam Nai talks about the *Yuon* he so loathes, his tone hardens noticeably. Mam Nai believes that the Revolution failed both because it was infiltrated by the enemy and because of the Vietnamese invaders. His language skills earned him the responsibility of overseeing the interrogation of Vietnamese prisoners, which took place in a house east of the prison, near a sewage canal.

"Did you believe their confessions?" asks Judge Cartwright.

"I think the soldiers made true confessions, because they were attacking us."

The judge pulls out a confession by one of those enemy soldiers. On December 14, 1978, the soldier "confessed" that the Vietnamese army was in complete disarray, that it would never have the courage to take on the glorious forces of Democratic Kampuchea, and that it was on the run from the Khmer Rouge. Three weeks later, Vietnamese troops took Phnom Penh.

"Did you believe this?"

> *What he said had nothing to do with the* Yuon *army. He was talking about one small unit. In general, I don't think it was true. Concerning the Cambodian prisoners, I don't think 100 percent of them were enemies. Nor do I believe that 100 percent of them were innocent. To some degree, some committed offenses. As for the Vietnamese, I firmly believe that they were the invaders of Democratic Kampuchea. None of them were innocent.*

**THE KHMER ROUGE CALLED** the Vietnamese "territory-eaters." In his S-21 diary, Mam Nai wrote: "We must make war today. And we must make war tomorrow. We must make permanent war. Will the Vietnamese succeed in eating us? That's up to us. If we can defend our country, we will be famous throughout the universe."

In Cambodian politics, it is always easy to hate the Vietnamese. Nowadays as much as in the past, Cambodian politicians need only raise the specter of the Vietnamese invader to win an argument, position themselves as patriots, or acquire some legitimacy. In 1970, the far right, led by General Lon Nol, justified its putsch by positioning it within the context of the national struggle against the "hereditary Vietnamese enemy." Later, largely to distract people from its own colossal economic failure and rule of terror, the Khmer Rouge was quick to assert that it was defending the nation from an aggressive neighbor. The party in power today got there on the backs of Vietnamese tanks, so it avoids that stance. The opposition, however, regularly turns to populist xenophobia to justify its own existence.

Despite their crimes, the Khmer Rouge are commonly described as "good nationalists." Far-right nationalism is toxic, whereas the far-left variety is honorable. The Khmer Rouge annihilated a quarter of the population of Cambodia, yet people say without batting an eye that the regime was devoted to its country. The Khmer Rouge loved Cambodia to death. Nate Thayer, the journalist who accompanied Nic Dunlop on his search for Duch, wrote of Pol Pot: "I think he was a true nationalist as well as a truly evil man." But is this a paradox or a combination of two evils?

A French journalist specializing in Southeast Asia wrote that nationalism is a way to fill the void created when Marxism meets reality. Prior to their victory, the Khmer Rouge leaders proclaimed that one had to choose between the individual and the Party, and that they had chosen the Party. Forty years later, in an interview he gave shortly before his arrest, Brother Number Two explained that one now had to choose between the nation and the individual, and that he had chosen the nation. At least he showed consistency in choosing the worse of the two.

# CHAPTER 29

**T**HE GHOST OF PHUNG TON HAS BEEN HAUNTING THE TRIAL from the start. Phung Ton, a law professor and dean at the University of Phnom Penh, is still remembered with great respect in Cambodia's academic circles. An important figure of Cambodia's liberal left, the professor was an expert in international and maritime law and was known for his progressive ideas and for leading a self-disciplined life in the service of knowledge and research. In 1968, the same year Duch was imprisoned for having links with the clandestine Communist Party, Sihanouk's secret police arrested Professor Phung Ton, held him for a month, and then placed him under temporary surveillance.

The professor had more luck when the Khmer Rouge took Phnom Penh in April 1975: he had left Cambodia under attack the previous month to attend a conference in Switzerland. He missed the brutal exodus from the city out to the co-ops, the forced labor in the rice paddies and on the canals and dykes; he escaped the tyranny of the young guards dressed in black and the ignominy of being "reeducated." But his wife and seven children were left behind in Cambodia. Time passed; Phung Ton had no news of them; the separation became intolerable. In a letter to a friend, a professor in France, where Phung Ton had taken refuge, he wrote:

> *I have a large family and cannot just abandon them to their fate.*
> *Let the new leaders in power put me in prison or kill me; no matter, so*
> *long as I see my wife and children again.*

On December 23, 1975, he announced that he was flying back to Phnom Penh via Beijing:

> *I don't know what to expect back home, but I want to go back if only to be reunited with my family, from whom I've had no news in nine months.*

On Christmas day, Phung Ton—along with a number of other Cambodian intellectuals, including Chao Seng, another renowned professor—landed in Phnom Penh. They had barely set foot on the ground when they were taken off to different camps. One year later, on December 12, 1976, Phung Ton was transferred to the prison from which there was no return: S-21. In the photo taken of him upon his arrival, he has the number 17 hanging around his neck like some kind of poacher's snare. He's wearing a shirt with thin stripes. He must have been the seventeenth person to have been registered at S-21 that month.

"My father looks unrecognizable and emaciated. His eyes are empty," says his daughter, showing the court the photo, the last visual trace of her father's existence.

It's estimated that a prisoner at S-21 lasted two months on average before being executed. The professor was held in his cell for at least seven. On July 6, 1977, a medical examination sheet, or at least what passed for one, recorded that Phung Ton was unresponsive, that he was suffering from diarrhea and renal failure, and that he was underweight. After that, his name disappears from the archives. His family believes that he died the day after the medical report was made. There's no record of his interrogations during the preceding months. A mere four sheets of paper concerning him have been found in the center's archives, constituting the first draft of a "biography." Those pages were written by Mam Nai.

DUCH'S PROBLEM ISN'T REPUDIATING the Party or Communism, or expressing regret for the purges or the merciless discipline, or reconciling with the three survivors of S-21. His problem is getting over his betrayal. Professor Phung Ton is one of a handful of individuals (Professor Chao Seng is another) whose executions at S-21 Duch finds especially troubling. The man has grown an impenetrable plate of armor against human emotion, yet the mere mention of Phung Ton's name can pierce it in a fraction of a second. To purge members of the army or members of the Party in the name of discipline or loyalty is one thing; but how does one justify the murder of a much-admired and harmless university professor, one who had the courage of his progressive convictions and whose only sin was to come home to his family? Duch struggles with it, even in the context of class warfare. The wound left by the murder of Phung Ton was opened at the very start of the trial and hasn't stopped bleeding, to the point that the professor seems to have become the victim around which Duch's fate is to be symbolically decided. Phung Ton represents everything we demand to know about this period, and on which hangs the possibility for the families to forgive Duch, as well as his impossible redemption. The professor's fate was the result not of violent, fratricidal purges but of a murderous ideology. That his daughter and widow show up in court every day without fail turns his fate into an emblem of Cambodia's tragic past.

**IT'S JUNE 16, 2009.** The trial has been going for two and a half months. Duch has spent many intense hours on the stand, testifying about the years before S-21, about Communist Party policies, about the establishment of the prison. Always self-possessed, and often in control of proceedings, Duch shows remarkable fortitude up until the morning when the court discusses the fate of Professor Phung Ton.

Duch starts to sob, the emotion twisting the features of his face. He tries to fight it, tries to fend off these waves of emotion that he usually keeps at bay, but they assail him like seawater pounding a sandstone cliff. Only when the presiding judge asks a series of confused and disjointed questions does Duch get some breathing space. Roux is on the edge of his seat. He doesn't take his eyes off of his anguished client.

"We—and this is particularly true in Western thinking—we expect those who have committed crimes against humanity to manifest some kind of culpability," says the psychologist. "There are various ways it can express itself: as a very serious depression, for example, or suicidal tendencies, or tears. Many show no guilt. But for those that do, this is how it comes out."

All the civil parties have left the courtroom. Duch tries to keep his answers to the bare minimum: "That's wrong," "That's not true." He knows that only by economizing his words can he stop himself from falling into the abyss of emotion that he has so carefully kept at bay for four decades. His voice has changed and he sounds slightly hoarse. The man who could modulate his voice at will now has trouble speaking. The man who dominated proceedings, who controlled, accepted, or rejected such and such an argument with irritating but masterful authority is now struggling to prevent his chest from being ripped open. His lawyer, Maître Roux, keeps his eyes fixed on his client. The lawyer remains perfectly still, his hands joined together under his chin and half-covering his mouth, as tense as a lookout atop a mast. He looks as though he's projecting all his energy toward his wretched client. The questions no longer matter; the facts, the details about the torture, all that has become background noise. Duch is drowning.

But he makes it through.

It takes him forty minutes to regain his composure, and even then, he's still fragile. In the afternoon, the court raises the issue of the medical experiments carried out at S-21. Duch's subordinates, Suor Thi and Prak Khan, talked about them in court. There are mentions of them in the archives. Yet Duch disputes at least some of the facts. Presiding Judge Nil Nonn asks the defendant if he was aware of these experiments.

"Yes, I was," says Duch, his voice calm.

> *Living prisoners were used for surgical training. Blood was taken from prisoners. My position regarding these blood transfusions has changed. During the preliminary inquiry, I said that the transfusions*

*were a vestige of Nath's time, and that I didn't know about them. But I*
*gave it some thought and remembered receiving a phone call from one of*
*my superiors, who told me that the blood transfusions were causing skin*
*irritations to combatants. This is another criminal act I committed.*

Duch has composed himself, but his conscience gives up some
ground. A hundred detainees died drained of their blood, he says. The
practice only ended after the medical staff was purged; that is, when
the only people competent to carry out this lethal bloodletting were
killed themselves. No one had been trained to replace them.

Pharmaceutical drugs were also tested on the prisoners. Duch
maintains that the prisoners were aware that they were participating
in an experiment. He admits to having been personally involved. But
then he tells a preposterous story: he claims to have secretly opened
dozens of capsules containing the test medicine, thrown it out,
cleaned the capsules with cotton buds and replaced their contents
with acetaminophen, thus saving his human guinea pigs from death,
he recounts to an incredulous and repulsed audience.

**"ACCEPTANCE IS A PROCESS,"** says the psychologist in court.

*The defendant goes through different phases. First, there might be*
*outright denial, a willful refusal to accept something. Next, there might*
*be some refutation, by which I mean the subject accepts some assertions*
*and rejects others. That leads the subject toward self-deprecation—in*
*this case very quickly, as we've seen. Duch immediately incriminated*
*himself: "I am a criminal, I feel guilty, what can I do?" I believe this*
*could well be another way of not accepting, of being unable to fully*
*accept the facts.*

"In this process from denial to self-incrimination, can certain obstacles remain? Things that are so difficult to bear that the subject
cannot bring himself to say them out loud?" asks Roux.

*Yes, that can be part of the process. The process can take a long time or just a little. It depends on the subject's background and on the thing that might lead him to awareness and acceptance. The trial is one contributing factor. We should take into account another: the saliency of the Khmer Rouge fabrications, which are still present. From an intra-psychic perspective, one might imagine Duch's conscience as a battlefield with different battalions moving across it: it is a mechanism with moving parts.*

Some parts of Duch's conscience shifted on that day in June. His confession no longer holds water. The man seems exhausted, spent after a long, frenzied journey from confession to confession, from setback to defeat. The morning's discussion of Professor Phung Ton left him shattered; in the afternoon, he confessed to a crime that he had never before acknowledged then talked wildly about the unlikely covert steps he took to stop the medical experiments on prisoners. His posture and his tone of voice have changed. He'd never looked as energetic and confident as he had the previous week; now his face looks haggard. "Heavy. Heavy and serious," Roux confides to me that night.

One name threw Duch off his game: Phung Ton. The trial's mysterious, uncontrollable force seems to be at work.

Yet it won't happen again. There will be other times during the trial when Duch will be confronted by his own emotions or by the truth, but he will never again let himself fall apart the way he did today. He won't make any further admissions; he'll go back to being master of his own confessions.

**AGAINST ALL THE EVIDENCE,** Duch denies knowing that the professor was a prisoner at S-21.

"I respected him. Had I known he was there, I would have given him my support, even if he was supposed to be smashed, since that was the procedure. I didn't betray the professor's soul, and I ask for forgiveness for his soul," he says, falling apart.

The professor's daughter sits ramrod-straight, a red scarf wrapped around her neck, and looks unblinkingly at Duch. Duch is adamant that Phung Ton wasn't tortured and that he died of an illness, not a blow to the back of the head and a knife across his throat. Yet when Him Huy takes the stand, there's a tremor in Duch's voice as he implores his former henchman: "You know what happened to the professor. Was he executed at S-21 or at Choeung Ek? Please, be honest . . ."

But, of course, it's hopeless. For Him Huy, a young soldier with no education, Phung Ton was just a number or a name on a list on which there were so many to check off.

There's little doubt Mam Nai knows more than anyone. He was in charge of interrogating the professor, who had been his mentor. Even worse: the professor's father-in-law, the dean of the Pedagogical Institute, had taken the young Mam Nai under his wing.

The prosecutor is the first to try to learn more about Phung Ton's fate. Without difficulty and with impeccable poise, Mam Nai immediately boxes him into a corner: "I don't remember interrogating Phung Ton. It's not coming back to me."

Silke Studzinsky, the lawyer representing the professor's family, doesn't do any better.

"I knew him, but I don't remember having interrogated him at S-21. I just don't remember" is the old Communist's unwavering reply. If the archives hadn't been kept so meticulously, or if they had been destroyed as they had been elsewhere, then Mam Nai's answers would be the end of it. The family would never know where the professor disappeared. Duch and Mam Nai could've denied the professor was ever even at S-21.

But there's that photo and those four sheets of paper with Mam Nai's writing on them. Faced with irrefutable evidence, Mam Nai grudgingly acknowledges that he drafted the professor's confession. Someone reads a paragraph out loud. Duch, a hand over his chin, looks at the screen on which the document is displayed. The conclusion reached by Mam Nai at the time was that the professor agreed with Communism. Now he has nothing to say about it. He reminds the court that once a person had been arrested and sent to S-21, that person had to disappear.

The defense lawyers manage to get a bit more out of him. With much effort, Mam Nai finally admits: "I carried out the interrogation. But no one made him confess. It was the same with Professor Chao Seng. They both spoke from their hearts."

"Thirty years later, how do you view that period in your life?" asks Roux.

"Can you be more specific? Are you talking about the current regime, or the previous one?"

"What do you think of Democratic Kampuchea?"

"Back then, there was nothing to eat. That was because of the war. But there were also positives: independence, self-control, self-sufficiency."

"Do you know how many people were killed at S-21?"

"My job and my position meant that I couldn't know that kind of thing."

"Do you know how many people died in Democratic Kampuchea?"

"I know even less about that. I just don't know."

"Do you regret having been an interrogator at S-21?"

"I'm not sure what you mean."

"Do you have any regrets today about having been an interrogator at S-21?"

"Yes, I have regrets."

"Can you tell us about them?"

"I believe that there were good people at S-21, and that wrongdoing took place there. But from what I saw, there were fewer good people than bad. I have regrets for that small group of good people."

"You have no regrets for the 'less good' people who were smashed?"

"I have never regretted the deaths of bad people."

# CHAPTER 30

**I**T'S A DICEY SITUATION FOR DUCH. IF HE STANDS UP FOR MAM Nai, it could be held against him; people might not believe he's telling the truth, despite his promise. If he sells out his former colleague, on the other hand, he betrays a loyal subordinate who, up to this point, he has appeared eager to protect.

Mam Nai and Duch both studied at the Pedagogical Institute in Phnom Penh when Professor Son Sen was its director. Mam Nai studied mathematics, physics, and natural sciences, and went on to become deputy head teacher at a high school. He discovered the Revolution in Mao's China. After spending two years in Prey Sar prison with Pon and Duch, Mam Nai returned to teaching, while his two comrades went into the *maquis* and set up the M-13 prison camp. In 1973, the repression of dissident teachers by General Lon Nol's military dictatorship drove Mam Nai and others to join the rebels. Duch had a job waiting for him at M-13. In 1975, the three former teachers became part of the team responsible for establishing S-21.

Within every mass crime lurk infighting and long-standing enmities, backstabbing and endless struggles between insiders and outsiders. Mam Nai describes how he lost Duch's full confidence around 1976, after a prisoner at S-21 implicated Mam Nai in counterrevolutionary activities. Duch made no exception for his colleague: he passed on the incriminating information to his superiors and let Mam Nai know what he had done. In the end, Son Sen let Mam Nai off the hook, saying he was a trustworthy intellectual. But the witness says Duch stopped trusting him with sensitive cases after that. From then on, Pon was in charge of interrogating important Party cadres.

Duch freely admits that he preferred Pon, one of his earliest comrades and a merciless interrogator. He puts it bluntly: Mam Nai was "slower" than Pon.

So, in fact, Duch hasn't always protected Mam Nai. Yet in court, he's quick to defend his former lieutenant. He maintains, for instance, that Mam Nai took no pleasure in torture. And whenever a problem arises with some document or other, the former prison director always manages to find an excuse. Mam Nai can't remember a list bearing his signature? Duch "understands" his subordinate's surprise. Someone else must have signed Mam Nai's name without him knowing it; the list's real author was Comrade Hor, who is dead, says Duch. Judge Lavergne gets peevish.

"I don't understand your explanation. What are you trying to hide?"

Duch offers no clarification.

"Fine. If these muddled explanations are the best you can do, we'll stop here. You may sit down," says the judge.

Duch's support for his former colleague has had a patently negative effect on the court. So when he takes the floor, he changes his tack completely. He gets to his feet and, snapping his arm in his brusque, martial way, berates Mam Nai and urges him to tell everything he knows about the fate of Professor Phung Ton.

> Please, don't be frightened by death. Just tell the truth. We stand here today before History. You cannot hide an elephant carcass under a basket. That's enough! Don't even try! I'm prepared to accept responsibility and answer for all the crimes I've committed. I want you to do the same. Please, remember that the civil parties are here and they want to know where our professor died. I think it is right that we should help them find the place. I don't think that we should let Communism live in our minds and stop us from telling the truth.

**NO ONE WOULD DESCRIBE** Silke Studzinsky as easygoing. The counsel for the civil parties makes no effort to please. "Silke," as everyone calls her, has a wiry frame, disheveled brown hair, and kohl-lined eyes; her robe is devoid of the white collar favored by the more showy lawyers. She glares at the defendant or at a hostile witness until the whites leap from her eyes like blades. Standing alone against the world doesn't bother her in the slightest. She seems quite impervious to the reactions from the outside. She's a rebel. She can be insulting and rude in court and in meetings; the wall she is fighting to tear down is a symbol, in her eyes, of a world unwilling to carry out justice. For a court lawyer to question a decision made by the president of the tribunal is considered the height of presumptuousness. Yet Studzinsky does so openly and bluntly, and more than once. In her own way, this unconventional lady is as absolutist as anyone.

It took just one quick question from Studzinsky at the start of the trial for Duch to get the measure of his opponent. His lawyer, Roux, offered some protection by quickly asking the judges to call Studzinsky to order; the presiding judge, though he hadn't yet taken control of proceedings, reassured Duch that he understood the need for courtesy. Duch was quick to adapt to Studzinsky's aggression. Whenever it was her turn to speak, the lawyer for the civil parties did so with the haste of someone worried that the victim lying on the ground has only moments to live.

"Do you acknowledge that M-13 was a killing center?" she says without offering any greeting.

"Good morning, counsel," says the defendant, eager to slow her down. "M-13 was a criminal operation."

"Can it be called a killing center? And please, keep your answers brief and precise," says the lawyer with a degree of venom to which she appears oblivious.

"I have no objection to that."

"Please answer succinctly, with a 'yes' or a 'no.' Did you explain to the children what kind of work they were going to do?"

She doesn't like Duch's answer.

"Could you please listen carefully to the question: what did you tell the children?"

"You're not listening to my answers."

"Did you tell them that they would be working in a killing center?"

"We didn't use that term. My duty was to teach them the Party line. They'd been called up to work for the Revolution. What did working for the Revolution mean? At that time, it meant killing people."

In theory, at least, the court forbids its participants from asking the same questions over and over, and Duch points out that those posed by Studzinsky have already been asked. He's pleasant enough at first. But eventually he loses his temper, raises his voice, and tells the lawyer not to interrupt him. "Let me remind you that I have the right to remain silent," he warns her. "That question has already been asked many times and I wish to remain silent," he says firmly. But the lawyer, stubborn as a goat, never lets up, even when she puts herself on the wrong side of the judges.

Studzinsky came to Phnom Penh to work for a German aid agency. She was the first Western lawyer to take an interest in the representation provided for victims at the trial, at a time when no one involved could care less about them. She met her clients, listened to what they had to say, and spent time consulting with her Cambodian colleagues. Her devotion to the cause is beyond reproach. But when she has to deal with the many foreign lawyers who are in Phnom Penh only for the short term, or with members of the court, she's as abrasive as sandpaper. It's as though she's at war with the world without knowing it. That Duch is running the show in court and imposing his version of history leaves her feeling as outraged as her clients, she says. In speaking so openly, she tells me, she feels "honest."

Psychologists tell us that each person is the product of both an individual past and a collective one. Lawyers who work at international tribunals tend to develop characteristics typical of stateless people. In the composite and amorphous international legal community, your home country is a big part of your identity. Accordingly, during her closing speech, Studzinsky introduces herself as a "German who lives every day of her life with the memory of her country's brutal crime." It's not the whole story, but it does give her a narrative at least. Her parents moved from East to West Germany six weeks before the Berlin Wall was built. She was seven months old at the time. Silke Studzinsky is no militant. Her background and her upbringing mean she dodged the siren songs of European totalitarianisms. But she seems to carry in her an obsession with the preceding generation's violence. For most of the trial, the bitter rancor she displays in the courtroom does her no good whatsoever and sometimes even does a disservice to her clients. Then, sometimes, this same unyielding vigilance saves her. She extracts from Suor Thi, the registrar of death, explanations that only he can provide. The explanations shed light on conflicting details in the detention documents related to Professor Phung Ton. Facing Mam Nai, Studzinsky has her moment of grace.

"Remember that the civil parties are here and they want to know where our professor died!" shouts Duch, hammering the point for his former subordinate.

The defendant hasn't yet had time to sit down when Studzinsky launches her attack. She immediately grasps the opportunity that Duch has given her and urges that Mam Nai be given one last chance to reveal what he knows. Her quick reaction is powerful. In front of the professor's widow, who is holding a handkerchief over her mouth, the former chief interrogator appears unsettled.

"I wish to express my regrets to the family of Professor Phung Ton," Mam Nai says.

Mam Nai the denier, the doctrinaire hardliner, cracks. His body starts to shudder and he breaks down in sobs.

> I was enormously remorseful because my brothers and my parents suffered and died under the regime. My wife died. I think that it was a chaotic situation and that we can only be filled with regret. It's the only thing we can do. Now, I try to find solace by thinking about my karma; I turn to my religious faith. Of course I

*had many regrets and I believe that through this court, the family of Professor Phung Ton is now aware of my feelings.*

Duch opened the breach in his former comrade's defenses. But it was Silke Studzinsky who seized the decisive, fleeting moment in which good professional reflexes can swing a trial around and split open a man's shell. Mam Nai's moment—painful, brief, and human though it is—will long be remembered by many of us who watched the trial.

When Studzinsky is finished, presiding judge Nil Nonn tries to push Mam Nai further. But the witness has already retreated into himself. He says that delving any deeper into that particular corner of his conscience would be like "shooting into the black night." That's all there is. Phung Ton's family won't get any new information about the circumstances of the professor's interrogation or death. But for a minute or two, for the first and last time, the curt, provocative Mam Nai broke down. The survivor Bou Meng, who was harassed by Mam Nai, rubs his forehead. Leaving the courtroom, Mam Nai passes within a few feet of the professor's daughter. She looks at him. He doesn't look at her. He walks stiffly away, an old and disgraced soldier.

# CHAPTER 31

**S**ON SEN, DUCH, MAM NAI, PON: THE MEN WHO RAN M-13 AND founded S-21 were all teachers. Revolutions, particularly those carried out on behalf of the "proletariat," are always imagined, willed, and led by intellectuals. To those betrayals that tear families apart—the brother who denounces his sister-in-law, the child who denounces his father, the wife who denounces her husband—can be added the betrayal of the elite by its own members, intellectuals denouncing their peers.

In the early '70s, the Cambodian elite—that is, the aristocracy and those who passed their baccalaureates and had access to university education—was so small that its members frequently crossed paths. In court, the victims' families often describe this privileged class's self-destruction, its members crushed by a revolution that they either couldn't or wouldn't flee in time, or which they embraced only to then be suffocated by it.

When Professor Phung Ton's widow walks to the witness box, Duch stands. She greets everyone except him. He waits for her to sit before taking his seat. She embodies everything that the Revolution tried to eradicate: an educated, elegant, urban teachers' daughter who worked in the civil service of the *ancien régime*. She represents everything Duch destroyed and everything that he most wants to be loved by. She calls the Khmer Rouge the "black-clad regime." Those three words contain all her horror of the Revolution. Phung Ton's widow feels no need to embark on a historical analysis of the legitimacy of Marxism; she harbors no naive nostalgia for some pseudo-egalitarian society. What she has is the kind of common-sense wisdom that comes from observing things with her own eyes: a regime that forces all women to get the same haircut and everyone to wear the same monochrome outfit can lead to only one thing: the end of happiness. If we were to reduce Pol Pot's ideology to just his fashion sense, any talk about the glorious future he trumpeted would fall apart. Fashion according to Pol Pot spells misery without end. Idealistic and well-intentioned people continue to debate Communist ideology, but more pragmatic minds who give such rhetoric and Maoist casuistry short shrift merely have to look at the Khmer Rouge uniform to see the grim future promised by Communism. For Phung Ton's wife, the clothes do indeed maketh the man. The regime's idea of itself couldn't have been clearer: it was the black-clad regime.

When her father fell ill at the forced-labor camp where they had been sent, Phung Ton's widow was obliged to address a young guard with a show of respect Cambodians usually reserve for their elders. The guard shrugged her off with a revolutionary slogan: there was nothing to gain from keeping the old man, and nothing to lose from getting rid of him.

"That isn't what Cambodian culture is about," says the seventy-year-old lady.

Her father died. Then one of her seven children died. Then an uncle. Then an aunt. One day, she brought up the idea of eating a chicken. The young, black-clad guard told her not to talk about such "bourgeois food." One of her sons was good at catching fish and tortoises in the swamp. He was punished for misconduct. Phung Ton's wife was accused of being a liberal city-dweller. They told her to "construct herself."

"I didn't know what 'construct yourself' meant."

**ALL THIS TOOK PLACE** thirty years ago, she says, but the suffering has only increased since then. "I have never been happy. I survive only because I'm on medication. Some people think I'm here for revenge.

That's not true. I'm here to see that justice is done for my husband, and to hear the truth: why did they commit such barbaric acts?"

The court takes a recess. Duch talks with a member of his team. A moment later, he's sitting alone on the defense bench. He looks across the room. The only person still there is Professor Phung Ton's daughter, reading her notes. She doesn't look at him. Eventually, Duch's lawyer returns, ending the defendant's solitude. Though she sits on her own, unlike Duch, the professor's daughter never seems alone.

Wearing a green blouse and dark, Western-style trousers and a jacket, Phung Ton's daughter takes the stand. Duch doesn't stand up for her. When the Khmer Rouge forced her to leave Phnom Penh and work in the fields, they also forced her to write her "biography." Rashly, she hid nothing, hoping that her candor would help her father find them. She didn't realize that such a "biography" could condemn its writer to death. By some good fortune, the Khmer Rouge didn't make her pay for her honesty.

What few family photographs have survived appear on the TV screens. There's one of Phung Ton and his wife on their wedding day, by the Seine in Paris; one of their seven children; one of the professor in his dean's office at the Royal University; another of him with two friends at a campground; and the last one ever taken of him, with the infamous card hung around his neck. Number 17.

The professor's daughter has been waiting for this moment for years. She and her mother have sat in the gallery every day since the trial began. They and Chum Mey are the only ones who haven't missed a day. And yet, even after having thought about it for years and years, when the time comes for her to testify, the clarity of her thinking succumbs to the chaos of her emotions, and her story drifts.

It was in October 1979, while buying palm sugar wrapped in newspaper, that she and her mother learned that her father had been killed at S-21. The newspaper had reprinted the photo of Phung Ton with the number 17 around his neck, which had been found in the prison's archives. The prison was already being converted into a genocide museum.

In the daughter's eyes, Duch is a "twisted" person who is still trying to shirk his responsibilities. Who decided to transfer her father to S-21? Who decided to kill him on July 6, 1977? What kind of torture was inflicted upon him? Duch says he can't answer the first two questions; as for the third one, he says that the professor couldn't have been tortured, since "torture didn't produce this kind of confession." But the professor's daughter is convinced that Duch knows how, when, and where her father died. A brave man would admit it, she opines.

Of the six individuals for whom Duch says he had the most respect before 1970, four were killed by the Khmer Rouge, two of them at S-21: professors Chao Seng and Phung Ton, both of whom returned from France on the same plane. Yet Duch is adamant that he had no idea Phung Ton was incarcerated in his prison. The face-to-face between the torturer and the victim's daughter ends in stalemate. With cold reserve, she looks into his eyes, then turns her back on him forever.

**THERE WERE SEVEN CHILDREN** in the Tioulong family, all girls. Their father was an eminent member of Phnom Penh's aristocracy. He was an ambassador, a minister, head of the government and head of the royal army. The family had close ties with Prince Sihanouk. Tioulong Raingsy was the second-born of that well-established family. She married Lim Kimari, an executive at the Cambodian Commercial Bank and a black belt at karate, very young. She had a job as a representative for a major Western laboratory and another as a presenter on a French-language radio station. When the civil war worsened after the coup of March 1970, the young couple sent their children to Paris, though they themselves stayed in Cambodia because, they said, they would have had a lower social status in France. They spent the summer of 1974 on holiday in France with the rest of their family, who had sought refuge there. At the end of the holiday, they made plans to return to France the following summer. A photo of Raingsy shows her soft, shiny hair and beautiful smile, though you can see the anxiety in her eyes. On March 28, 1975, from the besieged Cambodian capital, she wrote to her father:

> What advice do you give me? Should we leave as soon as possible? Or should we stay until June?

Maybe she decided the situation wasn't so bad after all: sure, the fearsome Communist guerrillas were on the verge of taking over; but Prince Sihanouk was also returning, and that was reassuring. Three weeks later, the Khmer Rouge entered Phnom Penh and the country was locked down. The window of opportunity was gone.

Like everyone else, the couple was forced out of the capital as soon as the revolutionaries entered it. They went to a village and lived normally until November. Raingsy and her husband were quintessential "new people," the sort that the Revolution handed over to the "old people," or "base people," for reeducation or execution. Revolutionaries spied on the couple in the evenings. They heard Raingsy speaking French, a death sentence in itself, and realized that she had no labor background. They interrogated her. She told them everything.

An S-21 document states that Tioulong Raingsy died on "April 31st" 1976. Clearly, the former math teacher's organization wasn't without flaw. April has never had so many days, not even at a time when each one felt interminable. The cause reads: "Beaten to death."

Tioulong Raingsy was tortured for months. In her parody of a confession, she said she was recruited by the CIA in 1969. Her missions included mobilizing the people to claim land, stealing water buffalo, and having a private life. She said that the network of traitors to which she belonged included Paul Amar, a journalist who, years later, had a successful career on French television. In the mountain of absurdity that is the S-21 archives, where the grotesque vies for pride of place with the horrific, his name is written "Pole Hamar." Just like the young sailor Kerry Hamill, Tioulong Raingsy mocked her tormentors by giving them names of people who were impossible to find.

At first, Raingsy's family in Paris wasn't too worried. French newspapers called the Khmer Rouge victory *la victoire rose* [the pink victory]. But with the victory came silence, and that silence grew heavier. It lasted four years, until the day when, after the Vietnamese had liberated the country, Raingsy's family learned about S-21. The archives contain Raingsy's interrogation documents, her photo, in which her hair looks ruined and the beauty is gone from her face, and a photo of her husband, Lim Kimari, who was executed one month after his wife.

What torments Raingsy's sister, who has come from France to testify before the court, is the guilt she feels over the confusion she imagines her sister and brother-in-law must have experienced when their family didn't come to their aid, and when the French weren't able to drive the Khmer Rouge from power.

The youngest of the seven sisters has come to speak on behalf of the family. Holding up the photo of her sister taken in 1974, she says she wants to show Duch what Tioulong Raingsy was like before. She wants to do this even if it makes no difference to him.

"I want to show him what he destroyed with his own hands."

The little sister doesn't believe the executioner's remorse. She thinks that, unlike his victims, he was lucky enough to get a fair trial. She says she will never forgive him.

"Never, never, never."

The children of the murdered couple grew up as best they could, she says. At the age of eleven, their son was diagnosed with neurological problems and epilepsy. Psychologists said that these were linked to the trauma he suffered. In the 1990s, he returned to live in Phnom Penh. One day he had a seizure while driving and was killed when he crashed into the Independence Monument roundabout downtown. One of his sisters also suffers from psychological issues. As for Raingsy's parents, theirs is a silent grief. Her mother still asks why the Khmer Rouge killed her daughter. In the early '90s, her father, who remained loyal to the king, had to negotiate peace with various parties, including the Khmer Rouge. He kept his feelings to himself.

# CHAPTER 32

**V**ICTIMS OF STATE-SPONSORED CRIME OFTEN FEEL AS THOUGH what they went through was the worst and most horrific crime imaginable. Though it's a pointless comparison that can only distress survivors of other crimes, victims tend to feel that their suffering is so great it can't be matched by that of others. For Tioulong Raingsy's little sister, "what happened under the Nazis happened here, except magnified, because here it was Khmer against Khmer." It was similar but worse "than that which the Nazis made the Jews endure" because, she says, the gas chambers were swifter. There's no reasoning with pain.

During the few days when the victims' families take the witness stand, there's no room left to reflect upon the banality of evil. To see the criminal as a man among men is a necessary scruple, the bitter fruit of appeased minds. But for those few days, that particular thought curls up and hibernates. Nothing resists the flood of grief, devastation, anger, disgust, indignation, and, sometimes, hatred. The victims' pain and wrath have us sinking deeper into our chairs like stakes being piled into soft ground. There's no question of forgiveness or reconciliation. There's no such thing as redemption. All that matters is punishing a torturer. A trial is an emotional dead end: when the defendant denies responsibility, the victims suffer; when he admits it, they suffer. Either way, they can't escape. To attend a trial is to experience a sort of asphyxiation.

Ou Windy was arrested as part of the same group as Tioulong Raingsy and Lim Kimari. They all came from the same privileged milieu. After the Khmer Rouge victory, they all found refuge in the same village. At thirty-one, Ou Windy was a civil servant in the ministry of foreign affairs, on secondment in the prime minister's cabinet. A married father of three, Windy was a graduate of the Cambodian National School of Administration and the first-born of a group of siblings who had all the advantages and were destined to succeed. His date of entry to S-21 is listed as February 13, 1976. His date of execution: May 20, 1976. His little brother has been thinking about it for thirty years. For 10,950 days and nights, he says.

> *I attended this trial because I wanted to try to experience what my brother went through; I wanted to share his suffering and fear, in my own way. I wanted to imagine the pain that you feel when someone hits you, when someone tears out your fingernails, when someone electrocutes you, when you're starving, when you're chained up. Your Honor, I thought I had no more tears. But I see now that I have more.*

The brother's testimony, given via satellite from Paris, falls at Duch's feet like a cast-iron block: a thud, followed by an endless echo. The dreadful tension only dissipates when the satellite link between Paris and Phnom Penh fails. The court examines a problem with the dates on the S-21 documents concerning Ou Windy. His entry date is later than the date on which his biography is registered. The victims' families turn over every detail in their minds the way a farmer tills every inch of his field. To them, everything matters, particularly details like these. Duch is asked to clarify the discrepancy. He has a perfectly coherent explanation for Ou Windy's little brother. "He would have been put in the special jail. That's why the entry date is different. The list of entry dates isn't always correct. The date of registration for his biography is more reliable."

Ou Windy's brother, like the parents of the victims, is haunted by what tortures he imagines might have been inflicted on his older brother. Yet he says he thinks Ou Windy's confession is written with "so much self-assurance and harmony" that he cannot see how it could have been written by someone who had just been tortured. The English-language interpreters muddle Duch's explanation for a moment, leaving everyone confused. Then he clarifies, in a clever understatement: "If I try to reassure Mr. Ou that there was no torture, people might think I'm avoiding the issue. I don't want to use this occasion to try to apologize for my crimes. So, to the extent that I'm not precise each time this subject comes up, I should like to repeat that torture was used only when it was unavoidable. The interrogators were all different. Some would quickly turn to torture, some wouldn't. I'm not going to say that torture wasn't used against your brother just to please you."

In 1992, Ou Windy's brother visited the haunted rooms at S-21. The youngest of the family, he is a successful businessman in Paris and a modest and reserved person. Like all Cambodians, he says, he keeps his emotions inside. At first glance, his expression looks severe. Yet his face lights up when he greets you, showing both warmth and reserve. There's something reassuring and calm about him, concealing his 10,950 nights of terror. In court, he describes his visit to the country:

> One evening, at a friend's house, I met a young woman who had a gift: she could communicate with spirits. I told her about my brother, and she contacted him. We Cambodians believe in this kind of thing. I don't know how to explain it. The young woman told me that she was in touch with my brother and that he was sad and terrified, that he had suffered greatly in the world of humans, and that he didn't want to be reincarnated. He said he was very frightened and that his soul had taken refuge in a pagoda; and he said that he had placed himself under the protection of the monks. The medium—I mean, the young woman—said the pagoda's name. The next day, I went to that pagoda with my sister and we held a ceremony. A very strange thing happened: inside the pagoda, I kept looking up at the ceiling and I felt that my brother's soul was there. So now, whenever I go into a pagoda, I'll look up at the ceiling, because perhaps my brother isn't the only soul who

*took refuge up there, to avoid reincarnation and ask for Buddha's protection. Now, my niece is pregnant. When the child is born, we shall go together to that pagoda, to introduce my brother to his grandson.*

**PARIS WAS STILL RECOVERING** from the student riots of 1968 when young Ouk Ket arrived on a scholarship to study engineering. He, too, was a member of Cambodia's privileged class; his family lived at the Royal Palace. After the 1970 coup, Ouk Ket answered Prince Sihanouk's rallying call and, quite naturally, joined the former sovereign's alliance, which included the Khmer Rouge guerrillas. That same year, he met a young Frenchwoman in Paris. They were married in October 1971, at the same time that, in a distant corner of the Cambodian forest, François Bizot was standing face-to-face with Duch.

A few months later, the young couple moved to Dakar, where Ouk Ket had been named third undersecretary at the embassy. Ouk Ket's wife gave birth to a son in 1973 and a daughter two years later. The young civil servant wasn't Khmer Rouge—his loyalty was to the king— but he appeared enthusiastic about the new regime he represented, whose dream of a better tomorrow he embraced.

*In Cambodia today, everything has been swept clean, everything is as clean in the city as in the country. There is complete security and guaranteed social equality. There is nobody on our backs exploiting us, and none of us shall be exploited. Therefore nobody will be rich and nobody poor. That is to say, it will be all for one and one for all. The factories will start working again, from the smallest workshops to the oil refineries. All the houses will be rebuilt, the schools reopened. Very soon, our children will have [a] radiant future.*

Ouk Ket wrote this to his father-in-law in December 1975, delighted to be "returning to a country benefiting from all this prosperity." In April 1977, a message from headquarters told him to return to Phnom Penh at last. The family spent three weeks in Paris. On June 7, still enthusiastic about the new regime, the diplomat flew via Beijing to the capital of Democratic Kampuchea.

"Ket was very happy to return to Cambodia to participate in national reconstruction," his wife says from the stand.

> *He seemed confident. On the bus, I was looking at his very handsome face when I intuitively said, "If one day someone comes to tell me that you're dead, I'll know that it will be because you've been murdered." He patted my cheek and said, "Cambodians aren't savages." Then he said, "Maybe I'll have to work in the fields a bit." That must have been the worst thing to him, I mean, the thing that seemed the hardest to him. Who goes back to their country knowing that they're going to be killed? He went home confidently, in high spirits.*

Ouk Ket sent a postcard from Pakistan and another from China, from where he wrote that he would land in Phnom Penh on June 11. After that, there was no more news. His wife heard nothing for two years. In December 1979, she asked the Cambodian representative to the United Nations for news of her husband. He told her, "Don't put your life on hold for him." Later, she learned about the existence of S-21, and that Ouk Ket's name was in the prison's archives. In 1991, in the middle of the peace negotiations then taking place under the aegis of the UN, she went to Cambodia for the first time, taking her two children with her. The family went to S-21 and to Choeung Ek. They searched the archives. On the forty-third line of a list of people executed on December 8, 1977, they read: "Ouk Ket, thirty-one years, Foreign Affairs, Third Undersecretary. Date of entry: June 15, 1977." He had been in cell 23, room 2, Building C.

Ouk Ket's widow describes how she decided then and there that the crime would not go unpunished. Usually, it's the victors rather than the victims who decide such things. But now, at long last, she can stand before the court and ask for justice, though of course nothing will ever satisfy that need. Neither Ouk Ket's widow nor his daughter refers to Duch by name from the stand. They refer to him as "Case Number One," which is what the tribunal designated his trial, its first case. Throughout their testimonies, the two women, in turn, use only this case number to refer to the man who reduced their husband and father to a number.

TIME RESOLVES NOTHING, particularly for the parents who come to testify before the tribunal. In the wake of any mass crime, there is always a small number of victims for whom speaking and condemning the perpetrators are vital processes. The vast remainder, including Ouk Ket's eldest son, stay silent. No one witnesses their suffering; nobody can sooth their enduring pain. For some victims, expressing their anger is a step along the path to healing. Yet that anger can seem like a river overflowing its banks. The need to talk about their suffering is endless; the story of their loss cannot be recounted too many times. Sometimes, the more they tell it, the sicker they become.

One rainy October day, I went to a provincial forum organized by the tribunal's office for the civil parties in its second case, in which the regime's four highest-ranking, still-living leaders were to be tried. The regional governor was to open the forum. She had hardly begun her speech when she burst into tears. Her father, husband, and son had all disappeared in Khmer Rouge "cooperatives." Her emotion was undiminished thirty years on. Then a Cambodian lawyer, only recently recruited to represent victims, declared that she, too, had been persecuted. She sobbed uncontrollably. Someone else admitted to having suffered psychological trouble and having had to consult specialists. An old Muslim man at the back of the hall got to his feet: "I am a victim of the Khmer Rouge. Is there a medicine to treat my mental problems?" Then, referring to the cases before the tribunal, he said: "We are dealing with only one germ. We all have all the other germs in our bodies."

Time doesn't resolve anything for Ouk Ket's widow, either. "For the past thirty-two years, Ket's absence has been unbearable. I miss him always," she says, looking up to try to stop her tears. "The pain hasn't faded; it has only gotten stronger. It's like an ocean in front of you. The result, for me, has been a complete breakdown."

Ouk Ket's daughter is older today than her father was when he died. She says that the day she put her finger on the S-21 register, a drop of poison entered her. Shortly after, she abandoned her studies. Like Kerry Hamill's younger brother, she ended up haunted by wild and uncontrollable thoughts.

"It was necessary for me to imagine it. Unfortunately, I imagined the worst."

When she found out that blood was taken from S-21 prisoners, she lost control. She sometimes feels as though she's the only survivor of all the children killed at S-21. Whenever she watches a ceiling fan spinning, she sees American bombers overhead, which, of course, she never actually saw. A deep sense of revulsion has taken root in her. When she describes it, her voice becomes cold, arrogant even, to help her hide her internal disintegration. She thought about suicide, about jumping from the window without knowing why. She reassures the court, says she's doing better. Yet a great sorrow hovers over Ouk Ket's daughter and wife. The older one gets, says the daughter, echoing her mother, the more the poison spreads. "The only way to return to my life is to testify."

# CHAPTER 33

UNLIKE TIOULONG RAINGSY, LIM KIMARI, OU WINDY, AND OUK Ket, Chum Narith wasn't born into Phnom Penh's upper class. He came from a background similar to Duch's, with whom he became friends. Narith's parents, though poor, wanted to give their children a good education. One of his younger brothers won a scholarship to study in France from 1960 to 1968. Narith received a similar offer, but the youngest boy was already in Paris and the family couldn't afford to send both. So Narith, the responsible one, turned down the scholarship. He became a teacher in 1965, like Duch. Mam Nai was one of his colleagues. Then, in 1968, again like Duch, and like Mam Nai, and like professors Phung Ton and Chao Seng, Chum Narith was arrested on suspicion of having links with the Communist guerrillas.

Chum Narith's younger brother is among the witnesses to testify before the tribunal. Circumstances in Cambodia forced him to become a French citizen before returning to Cambodia in 1999. By the end of the 1960s, he says, Cambodian society was already split into "blue Khmers" and "red Khmers." He draws a parallel between the situation in Cambodia and the uprisings then taking place in France. Cambodian intellectuals still had close ties with the former colonial power. They followed the events of 1968 closely and supported left-wing ideas. They opposed the social injustice that was widespread under the monarchy. Chum Narith's stint in prison only strengthened his political commitment and his opposition to the regime, just as their respective prison terms did for Duch, Pon, and Mam Nai.

From 1970 on, the civil war intensified and life became a lot harder. Refugees crowded into Phnom Penh, there were countless bombing raids, gas was scarce, and Cambodians went days on end without power. In 1973, there was an open revolt at the Pedagogical Institute. General Lon Nol's police believed Chum Narith was the ringleader and went to arrest him at his home, but he had already disappeared into the *maquis*, along with Mam Nai and several other teachers.

After the Khmer Rouge victory, Chum Narith joined its propaganda unit in Phnom Penh. On October 29, 1976, he was arrested, along with his younger brother, Sinareth, and Sinareth's wife, Dong Sovannary. It turned out that one of the teachers with whom he had gone into the *maquis* had later been arrested and sent to S-21. There, he was tortured by Pon—another former colleague from the national education system—before he denounced Chum Narith in his confession.

Chum Narith was accused of forming a group opposed to collectivization. At the trial, his younger brother asks the court how anyone can believe a confession obtained by torture. Yet despite this, he seems to *want* to believe that the charges against his older brother were true, as though Narith's admissible arrest by a regime founded on lies, fabrication, and slander could mitigate his powerlessness and rage.

Chum Narith was executed on January 1, 1977, after sixty-five days of prison and torture. His brother's voice swells until it fills the room: "I don't understand the point! If you want to kill, why not kill immediately?"

Chum Sinareth also died at S-21. The ignominious sign around his neck bore the number 59. His date of execution is unknown. His wife, wearing Khmer Rouge clothes and a Khmer Rouge haircut, was number 18. All that remains of her is a photo.

The youngest Chum brother happened to be in France in April 1975, and ever since he has been struggling with the sense of guilt so common among those who, by sheer luck or fortuitous circumstance, survived. He feels deep regret, he says, for not being intelligent enough, for not having the presence of mind, for misreading the situation, for failing to foresee the coming terror. But from Kigali to Phnom Penh, people never imagine the worst will actually happen, even when all the signs are there.

Where were his brothers executed and buried? The question still haunts him. It wasn't at Choeung Ek, which didn't exist yet when they were killed. He wrote to Duch, asking him, but Duch replied that he didn't know. The youngest brother doesn't believe that Duch had no choice but to follow orders. He thinks Duch enjoyed his work; he thinks Duch was a predator.

> In Christianity, there is the story of Cain. He killed his brother, but Abel's eyes followed him everywhere, to the point that he could never be at peace and had to ask someone to dig a hole and bury him in the earth. A French author once wrote: "And after they had shut the crypt upon his brow, / The eye was in the tomb and looked at Cain." So even though he was buried, his brother's eyes followed him into the grave; they followed the corpse. More than twelve thousand people died at S-21, which means twice that number of eyes. Twenty-four thousand eyes follow the defendant every day and ask him to explain. In Christianity, his sins are forgiven. But in Buddhism, good is rewarded with good. I believe that, right now, there are more than twenty-four thousand eyes following the defendant. There is nowhere he can go to hide from them.

During the recess following Chum's deposition, Duch stands and smiles. Members of the victims' families have been testifying one after another, each account proving tenser and darker than the previous one. For the former executioner, there's no way out. The sense of discomfort and contrition he showed early in the proceedings seems to have disappeared, as it has been continuously rebuffed by the victims, who communicate their mutual support to each other through the glass wall. The fierce and eloquent testimony given by Narith and Sinareth's brother has galvanized them.

Duch's Cambodian lawyer, Kar Savuth, wanders among the rows of civil parties. He talks to Phung Ton's widow. The professor's daughter joins the conversation, which appears cordial. Kar Savuth is of the same generation as the professor, and belonged to the same Cambodian elite of the 1960s. Phung Ton had been his law professor. Lon Nol's former minister of culture, whom the Americans evacuated on April 12, 1975, is in the audience. His elegant wife, who has all the poise and grace of the old, cultivated elite, had once been a student of the wife of Son Sen, head of Pol Pot's security apparatus. Son Sen's wife had once been a decent but strict woman, her former pupil tells me. Then she became a hard-nosed revolutionary and ended up devoured by the revolution she served.

If it weren't for all the adversities and betrayals, you'd think you were at a family reunion. Cambodia's elite constituted a small world in which everyone knew everyone else and in which, before the Revolution, everyone's path crossed everyone else's. The story being written during the course of the trial is like an explosion in midair: how Cambodia's intellectuals clung to the privileges that had allowed them to flourish even as the wings fell off and they found themselves hurtling toward the ground; how they allowed the flames of change to flicker to life among them, never imagining the conflagration to follow.

"Many old friends were imprisoned at S-21," says Duch.

> *Chum was among those I betrayed. I really had to keep away from them. I didn't want to see them. I couldn't face it. As for the twenty-four thousand eyes, I understand that thinking, and it's because of this that I accept that the civil parties point the finger at me. I am being very sincere at this moment. I feel compassion and I am filled with remorse. I honestly acknowledge and accept all the statements you have made.*

A little earlier during the trial, Duch pointed out that no steps were taken to alleviate the mental suffering of prisoners at S-21. The prisoners, he said, were considered no more than animals, or even less.

Had he distinguished between friends and strangers, he would have been accused of consorting with the enemy. This, he said, was the trap preventing him from showing even the slightest degree of empathy for the prisoners. Nor had he wanted to risk showing any emotion, he said.

Psychologists call this type of behavior "reaction formation." Duch resorted to blind obedience, overzealousness, and total allegiance in an "over-adaptation to terror" in order to suppress his own fear and silence his own doubts.

"How would you characterize this attitude of avoidance?" is how one of the judges poses the question to Duch.

"I don't know. I shut my eyes and ears. I didn't want to see reality," he says, his voice cracking slightly.

"Was it cowardice?"

"I think it was more than cowardice. I didn't go to see my friends at S-21 because I didn't know what to say to them. Certainly, I was a coward. But it goes further than that, because I betrayed my friends and teachers in order to survive. It was more than cowardice."

# CHAPTER 34

**D**UCH DOESN'T HAVE ANY PSYCHOLOGICAL DISORDERS. HE ISN'T suffering from any neurosis, psychosis, or psychopathy. We cannot comfort ourselves by dismissing as deviants those men and women who perpetrated mass crimes in extreme political circumstances. Duch is neither mentally ill nor a monster, and that's the problem. He wasn't dangerous before 1970. And he most likely wasn't dangerous after 1979. The same applies to Him Huy, Suor Thi, and Prak Khan. Duch will be punished for the rest of his days, because somebody must be punished. But he could be rehabilitated: the twenty years between the day S-21 was shut down and the day of his arrest prove it. What's more, no one has questioned whether it's safe to let Mam Nai and other former members of the S-21 staff—or the tens of thousands of former Khmer Rouge cadres who, for the past fifteen to thirty years, have been living freely alongside those they persecuted—to continue doing so.

A year and a half before the start of the trial, the S-21 survivor Chum Mey told me that he could see no reason why Duch should be released. He wanted and demanded a severe sentence for his torturer. Yet he also seemed more worried about Duch's safety than his own: "We are safe now. I wouldn't be frightened if he were released. It's the tribunal that should be worried: what will it do if he's killed?"

The psychologists have less trouble addressing political crime. Their training precludes them from believing that anyone is born a monster or devil.

"Whether they sponsor it or carry it out themselves, people aren't born torturers. They turn into them," says the court-appointed psychologist.

> *Every torturer who dehumanizes his victims was first dehumanized himself. This isn't an excuse but rather the key to understanding the psychology of someone who commits a crime against humanity. A person can be dehumanized by experiencing or witnessing cultural humiliation or personal humiliation. The person then does everything he can to compensate for those humiliations and disappointments, to the point of denying the humanity of the person or class of people he deems responsible. The person who commits crimes against humanity first eradicates his own individuality before denying it in others. Duch always falls back on reason, on logic, on mathematical models. He has literally smashed his own personal identity, if I can put it that way, in order to make room for the only kind of identity that matters to him: the collective kind.*

One night after the trial, I was eating dinner in a café with some acquaintances who were passing through Phnom Penh. One of them confidently declared: "Duch is a pervert."

That seemed to settle it for him. He had found an *explanation*. The denial, the manipulation, the effort to control others, the desire to please, the striving to impress people for his own self-benefit: aren't all these constituent parts of perversity in its broadest sense? And aren't they all blatantly obvious in Duch? The answer to both questions, of course, is yes. But, alas, to say that this explains Duch's actions is to mistake *observation* for *explanation*.

"We could talk about the notion of perversity," says the psychologist. "But then we'd simply have to work out where *that* comes from. Perversity doesn't exist in and of itself."

The perversity explanation might provide some intellectual comfort, but it resolves nothing.

We all develop "life strategies" with which to negotiate our inner contradictions and overcome the obstacles of life. Duch developed strategies first to serve the Khmer Rouge, then to survive it; his strategies included zealotry and compartmentalization. Today, he has psychological mechanisms that allow him to exist, or survive, both with and despite his crimes.

In order to understand Duch's actions, say the psychologists, we must examine the extreme way in which his collective and personal stories overlap. Duch's psychology cannot be separated from the society around him, or from the collective history in which he has been swept up. Duch doesn't exist *ex nihilo*. Both he and Cambodia went through "successive and massive acculturation," followed by a brutal and radical transformation into the New Man: that identity manufactured and demanded by the Khmer Rouge in which the individual exists only for the group, in an atmosphere of mistrust and generalized fear, and with all emotions and personal thought eradicated. Either you adapted or you died. In ethno-psychiatric terms, Communism constitutes a "deculturation." It flourishes in the uprooting.

"The man who lives in a country under totalitarian rule has a different psychology than the man who lives in a democratic state," says the expert.

Five months into the trial, the quality of the silence in the courtroom has changed. No longer is it that breathless and dumbstruck silence that knows it is watching history being written, nor is it the solemn quiet of a legal drama. The silence that fills the courtroom now is that of fatigue, of weariness, of exhaustion with both the trial and Duch's words. His performance has lost its shine. Now he sounds like he's rambling aimlessly. David Chandler, who has dedicated so many years to studying the tragedy of Democratic Kampuchea, and who has spent more time immersed deep in the S-21 archives than anyone else, has another way of measuring this decline: he believes the Khmer Rouge leadership's disconnect from reality, and the extent of the catastrophe that its reign brought about, points to its "profound stupidity."

Duch reveals the limits of his own intelligence and cunning. On the stand for a final cross-examination concerning his personality, he maintains his story; other than a few minor details and dates, he no longer has anything important to say that he hasn't already said many times over. He talks in tedious, ineffectual circles. He has lost his mental agility. He is a shadow of the man who, five months previously, had the upper hand. He no longer cares, and he no longer holds our attention—his luster is tarnished. After forty minutes, he ends his own deposition and lets the presiding judge take over. The trial isn't over and yet it's already over. We face the remaining six days the way a boat enters port: motoring slowly, sails furled. The last six days have been set aside to hear defense witnesses. Duch's mother, who is on the list, won't testify. On the penultimate day of the trial, the defendant reiterates his apology:

> I must bear responsibility for the crimes I committed. As I've repeatedly said, you can't hide an elephant under a rice basket. The enormity of the crimes committed can't be concealed beneath two leaves from the tamarind tree. I think that's all I need to say to the court, and that's the real truth.

Duch respects the two shrinks who examined him. Their work, he says, "is based on purely scientific, unbiased reasoning." His lawyer François Roux also praises the experts' psychological report; at the start of the trial, Roux had framed the issue thus: "Have you ever known a child who dreams of growing up to be an executioner?" From the first day of the trial, Roux warned the victims that his client would be unable to answer all their questions, and that there was no simple, unambiguous, comforting answer to the single question haunting the survivors and the families of those who perished: *Why?* The lawyer tried to prevent the victims from harboring false hopes. Yet in so doing, he laid bare his own: "Will we be up to the task of not only giving back to the victims their humanity, but of readmitting to the human race someone who has abandoned it? That is the great challenge facing our court."

At first, Duch keeps his lawyer's dream alive. He takes care to politely greet the judges, then the prosecution, then the civil parties. Walking past the long window separating the court from the gallery, he sees the survivors Chum Mey and Bou Meng, and salutes them. They smile and return his greeting. When the first lawyer representing the victims' families addresses him, Duch gives him his undivided attention. He strives to be cooperative, to give full and detailed answers. He is even more considerate with the next lawyer, who wants to know how Duch ranked different transgressions at S-21: "I will try to answer. If I don't, please ask me again."

Roux is candid about his own utopian vision for the trial:

> Duch has said, "I'm the one who gave the orders. I assume responsibility." There aren't many people in this country who have admitted that they gave the orders. Are we to say nothing about the fact that he also received orders? What took place under his command also took place above it. Do you think it's easy to come to this courtroom and say publicly: "I acknowledge this. I am ashamed of all I have done?" Do you think it's so straightforward? Duch has been on a long personal journey for many years. Who could have imagined that the once all-powerful director of S-21 would one day return to face its survivors and guards, flanked by two police officers and two judges? Who could have imagined that? Whatever the tragedy, let us appreciate for a moment that today, this man is confronting his past. It takes a certain amount of courage to do that. What is it that keeps him alive? He is convinced that he still has a role to play in the human race, and that that role is to ask for the victims' forgiveness. Duch is still a human being. Perhaps he struggles to admit certain things. But then, maybe you also find it difficult to admit certain things on his behalf. Anyone can make a mistake— even a prosecutor. And one can be mistaken in good faith when one stands accused. I dream that, by the end of this trial, the victims as well as the Cambodian public will be able to say that at the very least, now we have some peace of mind. If that happens, then we shall be able to say that justice has been done.

François Roux likes symbols. All his life, he has dreamed of a better world and has never stopped working toward it, whether this has meant standing alongside the peasants of Larzac fighting for their land in Southern France against the extension of a military zone; or conscientious objectors opposing the military draft; or Kanaks fighting for independence from France in New Caledonia; or just as well defending the rights of those accused of genocide against the Tutsis in Rwanda or representing a member of al-Qaeda tried in the U.S. for his role in the September 11, 2001, attacks—in the name of opposing the death penalty. The philosophy of nonviolence made a deep impression on him, and he has always been staunchly on the side of those who rise up against injustice. Yet at the same time, he has always respected the state and its institutions. Roux, who is descended from Huguenots, regularly returns to his rural community, surrounded by high mountain pastures and flocks of sheep, for rejuvenation. He is also an avid traveler, an idealist keenly aware of the great power of doubt but who finds even greater fulfillment pursuing the bright promise of a better world.

The international justice system is filled with symbols to fire the imaginations of human-rights activists, and Roux has found in it the building blocks of the better world he dreams of, though he knows that most of the time, it will remain just that: a dream. The dream that he carefully and passionately constructed in Duch's case centers on what he considers the most important event of this particular legal undertaking: the reconstruction that took place at the scene of the crime, one year before the start of the trial proper, when Duch returned to S-21 some thirty years after having left it.

That day, Duch stood facing the prison's three still-living survivors—Bou Meng, Vann Nath, and Chum Mey—and read his declaration:

> *I am completely overwhelmed to be in this most painful place for my countrymen and for myself. My first thought is for the victims and their families. They suffered innumerable miseries and inhuman*

*tortures and insults before dying. I feel a great and indescribable remorse, which I hope is made manifest by my accepting to stand trial alone for S-21. I am determined to do everything I can to bring justice to my compatriots, to the victims of S-21, and to their families.*

*I also feel enormous regret for all the S-21 cadres who were forced to carry out their tasks alongside me. That is to say, to carry out tasks they hated and that their parents hated, tasks to which some of them eventually fell victim. I feel great pain when I remember those events.*

*I sincerely regret having accepted the ideas of others, and having agreed to carry out those criminal tasks that were entrusted to me.*

*When I think about it, I realize that I am angry first and foremost at the Party's governing body, which did everything in its power to lead the movement to complete and absolute tragedy. Then I am angry with myself for having accepted the ideas of others, and for having blindly followed their criminal orders.*

Then Duch asked for forgiveness. He became confused, looked around, pursed his lips, opened his mouth, seemed to be on the verge of reading again, before turning and handing the sheet of paper to his Cambodian lawyer. He took the text back from the lawyer, put his spectacles back on, took them off, hesitated.

"Take a moment, catch your breath," said Roux quietly.

Duch was flushed, sweating, his glasses slipping down his nose. He started reading again, a quiver in his voice. Vann Nath sat facing him, with his head lowered and his arms crossed. His eyelids fluttered. Chum Mey sat less than two meters away, watching Duch. When Duch asked for forgiveness, Chum Mey nodded. "I know that my remorse, as painful as it is, is but a drop in the vast ocean of agonizing wretchedness felt by the victims and their families."

Vann Nath looked at his feet, hiked up his trousers, crossed his arms, kept his eyes fixed on the ground. Duch ended his solemn statement with a series of *sampeah*s, his hands pressed together in front of his face. Vann Nath sat up, his lips pursed, his arms still crossed over his chest, his gaze still turned downward. Chum Mey gave his torturer no more than a brief nod. Duch took off his spectacles, acknowledging his pain without hiding it. Chum Mey asked to speak. He rose and stood a meter or two in front of Duch. "What I want is freedom, and freedom is what I have now. I thank Duch for having agreed to come here, for testifying, and for recognizing his responsibility."

Duch pressed his hands together in gratitude. He gulped as he breathed, as though his breath were hanging by a thread.

"I have no rancor toward him," continued Chum Mey. "What I want is justice and peace for our country and for the million citizens who were killed. The only issue now is that he must speak the truth in court."

The former executioner's apology appeared to be coming to an end. Then Bou Meng stood up, his hands pressed together in front of his face. "I have heard Duch's declaration and I am 100 percent satisfied. I ask the judges to judge Duch according to national and international standards."

He sat down again.

Vann Nath didn't move.

THIS UNSETTLING MOMENT, which occurred during the trial's secretive investigative phase, was many things: extraordinary yet incomplete, moving yet affected, spontaneous yet staged. Most of all, it was the moment when Roux hoped a redemptive reconciliation would emerge from the horror.

But justice, like a revolution, is a graveyard of broken dreams. Two and a half months into the trial, François Roux realizes that there can be no *entente cordiale* with the prosecution. Something has been irreparably damaged in the trial the way he had imagined it, and when it draws to a close a few months later, Roux will be openly at war with the prosecutor. The portrait that the prosecutor paints of the defendant irritates Roux to the highest degree. With icy bitterness, he reminds the court of what the prosecutor said in his opening argument: the only logical conclusion that could be drawn from the facts was that "rather than someone acting against his will, practically unaware of the atrocities his subordinates were committing all around him while he remained holed up in his office keeping meticulous records, Duch, in fact, enjoyed the complete confidence of his superiors and was therefore responsible for carrying out at S-21, with great devotion and no pity, the Communist Party of Kampuchea's policy of persecution against the people of Cambodia." If—and only if—Duch admitted to this, declared the prosecutor, then he could claim to have confessed to his crimes and thus benefit from any concessions that such a confession might entail. His voice white-hot with anger, Roux addresses his client: "Duch, our arguments are drawing to an end. I have only one question for you: do you admit that, in reality, you enjoyed the complete confidence of your superiors, and that therefore you were responsible for carrying out at S-21, with great devotion and no pity, the Communist Party of Kampuchea's policy of persecution against the people of Cambodia? Do you admit that, yes or no?"

"Yes, I admit it completely."

"You went back to Choeung Ek and S-21. My question to you, Duch, and I'm speaking to the man now, is this: what did you feel when you arrived there on that morning in February 2008? Please, tell us what you felt in your heart. Talk to us, Duch!"

Duch can't hide his tension. He takes loud breaths between sentences and explains, in a voice devoid of emotion, how he was determined to go to these places to reflect, to apologize, and to ask for forgiveness from the souls of the dead. The silence in the courtroom could not be thicker. But Duch remains deep within his steel armor. He says no more. He doesn't break down. The carapace of control over his emotions proves more powerful than his lawyer's appeal for the emotional truth.

Roux is reduced to showing the video footage of that day in February 2008, during the famous reconstruction at the site of the crime. In the video, we see the Duch that Roux wants to show, who is real, not a sham, but it is only one aspect of Duch. The prosecutor opposes the viewing. The film contains a Duch he has decided never to see. In any other international court, a defendant who pleads guilty would get the prosecutor's support. No one questions the penitent's "sincerity" when he expresses his remorse, an act which is particularly valued to convince the judges. But this isn't the case in Phnom Penh, even though Duch alone has admitted to the bulk of his crimes.

Unlike at other international tribunals, the inquiry falls outside the prosecutor's control in Phnom Penh. Here, it is in the hands of the investigating judges. Consequently, there can be no direct negotiation between a cooperative defendant and the prosecution. The prosecutor's office has never really tried to understand this legal system, with which its top members are not familiar. Frustrated by it, they arrogantly dismiss it out of hand. As a result, the prosecutor sees his job in the narrowest terms possible—that is, prosecution and punishment and nothing more. So the Duch he sets about attacking in court is the powerful Khmer Rouge commander who ordered death with faith and zeal, who relished his role as a torturer, and who was proud of what he did on behalf of the Revolution. This Duch is no illusion, of course. But nor is it all there is to Duch.

François Roux tries one more time, against all odds, to guide the trial toward the conclusion he would have preferred. "Do you authorize me to say to the victims that, if they should so wish, they may visit you in your cell and that you will open both your door and your heart to them? The road doesn't end here today. It can continue between them and you, if they wish it."

"I would be very happy to receive any victim who wishes to meet me. I will open the door to them emotionally, and I would like the victims to finally acknowledge that I have accepted my responsibility and my guilt. So whoever you are, my door will always be open to you."

But a trial is not a dream, and good intentions carry little weight in court. A trial is a physical, violent act that elicits vehement reactions from its participants and observers. A trial drains souls, frays nerves, makes the pain worse. On September 17, 2009, on the eve of Pchum Ben, the Cambodian festival of the dead, Duch's trial ends. The parties must come together one last time at the end of November to make their closing statements. At this stage, all the frustration, pain, and anger cluster around the defense, because it was the defense that dominated the trial. After six months of intense and sometimes dramatic arguments, François Roux finds himself being virulently challenged, particularly by the victims' families. He feels a deep sense of failure: nobody accepts the Duch whom he defended. And the worst of it still lies ahead. Sometimes broken dreams are like revolutions: they turn into nightmares.

# CHAPTER 35

**E**VERY YEAR IN NOVEMBER, PHNOM PENH IS INVADED BY PEOPLE from the countryside. In 1975, the Khmer Rouge emptied the capital of its inhabitants, forcing the people of the city out into the country so that they would learn from the people of the fields what the New Man looked like. Pol Pot's partisans were not the first to come up with the idea of banishing city dwellers to the country. The verb "rusticate" used to mean to work in the fields or live in the country. But in the eighteenth century, Britons started using it to mean to forcibly send someone into the country—wayward students, for example, or people suffering from certain illnesses. In Revolutionary Cambodia, people were systematically rusticated en masse, until the end of their lives. The city was deemed a cancer of corruption and money, bad in itself and bad for all. Now, once a year, the peasants take their revenge. Thirty-five years after the people of the city invaded their lands, the people of the fields invade Phnom Penh and make it their own. For three days in November, the city puts on the huge street party that is the Water Festival. In other words, the peasants de-rusticate themselves.

Some people will tell you that the Water Festival celebrates a great naval victory that took place in the twelfth century, during the Angkorian golden age, but that's not what all the excitement is about. The festival takes place at a time of year when the Tonle Sap, the river on which Phnom Penh lies, reverses its flow. The Tonle Sap flows into the Mekong, and every rainy season when the Mekong floods, the Tonle Sap backs up. Then, when the flood subsides, there's a brief moment of slack water before the river resumes its natural flow down-stream. The festival takes place under a full moon, when the current has just switched direction. During the Water Festival, Cambodians give thanks to the majestic Mekong and to the mythical serpents for making the soil so fertile and the fish so abundant. The main event is the boat races, when traditional pirogues, with their very long, narrow, and unstable hulls, carrying up to eighty rowers each, line up in pairs to race along the river. Hundreds of thousands of mostly poor but always cheerful villagers converge on the riverbanks around the Royal Palace. It's at this time of year that a certain type of Phnom Penher likes to declare—preferably with a degree of trepidation—that the only thing to do is to get out of town before it's invaded by the army of beggars and rogues. So there's a two-way migration: the people of the city rusticate themselves and willingly abandon their metropolis to the people of the fields. The Tonle Sap, meanwhile, quietly hesitates over which direction to flow. At nightfall, when the crowd has thinned out and the remaining festival-goers idle along the promenade, a simple, jovial, and familial atmosphere takes hold of the capital, liberating it of the self-importance and hurry of the urbanites.

In 2009, the festival occurs early in November. By the time the closing statements begin, reuniting various members of the trial's broken family after a long separation, the Tonle Sap has finally switched direction and is now flowing toward its natural endpoint: the open sea. The monsoon is over. It's the start of the few short weeks of what passes for winter in these hot and wet latitudes. Cambodians begin to complain about the cold and pull out their sweaters and beanies, while the few thousand Westerners who live in the country quiver with excitement at the prospect of a short respite from the heat. And although winter in Cambodia feels like little more than a warm late spring in the northern hemisphere, it's endearing to see all the expats tantalized by the prospect of wearing jeans without feeling as though they're melting under the weight of a wetsuit, or the way they nonchalantly throw lightweight sweaters over their shoulders, each trying to appear more casually stylish than the next. Where they come from, their senses awaken as they shed layers in spring; here, they revel in being able to cover up for "winter," one which Cambodians actually feel. The light shimmers in winter, soft and sharp, a light washed by recent rains but not yet bleached bone-white by the heat to come. The sky is filled with velvety, electric blues and reds, colors that occur nowhere else in the world except, I'm told, in the Arctic north. Those popular paintings of the Far East, done in colors that once seemed so kitschy and artificial, now seem almost realistic to me.

In court, there's a sense of the magnitude of the occasion, and with it an indefinable atmosphere of anticipation, of emotion and of artifice. The players in the legal drama return to the stage one by one. Ouk Ket's wife and daughter have come from France, as has Ou Windy's brother. Tioulong Raingsy's sister is here. Professor Phung Ton's widow and daughter are present. There are unexpected people, too, such as a judge from the newly formed Special Tribunal for Lebanon, the most recent of the world's international tribunals. I'm told he has no fewer than seven bodyguards to travel around Phnom Penh. I remember that in Beirut, next to the mausoleum of the former Prime Minister Hariri, are the tombs of his bodyguards. Not one of their seven bodies was sufficient to shield him from the bomb that was destined for him.

Compared to their Lebanese colleague, the judges on the tribunal for the Khmer Rouge can be grateful for their lot. The danger he faces isn't one that hangs over them. Under the authoritarian, occasionally vulgar, and always shameless rule of Prime Minister Hun Sen, Cambodia can be a dangerous place for anyone trying to make it more democratic, or who would like to see its wealth distributed more equitably. But the country once ruled by Pol Pot is a haven of safety for this tribunal's judges; they have no need for the endless and futile hassle of the security detail. The world of international law can be both fearful and self-aggrandizing, and many of its top judges and prosecutors measure their own importance by the number of bodyguards attached to their persons. With one exception, those of the Phnom Penh court enjoy moving freely around the city on their own.

**IT'S TIME FOR THE CLOSING STATEMENTS.** For some people, it's also time to put in an appearance. Mr. Sur is such a person. When the curtain is lifted and the courtroom revealed, he is standing proudly in the middle, in front of the benches reserved for the civil parties, facing a public that perhaps he thinks is on his side. The French lawyer, who said that he would come only to the first and to the last days of this trial, has proven true to his word. His British colleague, Mr. Khan, stands beside him; we haven't seen him since the opening arguments, either. Mr. Sur and Mr. Khan greet one another like two cuckoos in another bird's nest.

Nearby, a young Cambodian lawyer is obliged to give up her seat in the first row to a European colleague; the European is older and more senior, and has just made a very long journey. Still, she's been sitting there throughout the entire trial. Few lawyers can resist pride—in France, lawyers are called "maître" (master) and meet in the justice "palace."

Another person to show up in court for the first time in a long time is the Cambodian co-prosecutor. Like her Canadian counterpart, she didn't want to expose herself in the public arena, and instead cravenly let junior lawyers—or, for that matter, anyone who wished to—take on the responsibility of prosecuting Duch. Consequently, at least five lawyers have been in charge of the prosecution over the course of the six-month trial. Which means that really, no one's been in charge. In another time, under different circumstances, lawyers would have considered the position a high point of their career, a professional honor. But in today's international tribunals, and in Phnom Penh most of all, public prosecutors tend to seek their glory more at international conferences and meetings, which are relatively safe, rather than in the courtroom, which is not. Each generation has its own idea of prestige and panache.

The dubious honor of giving the first closing statement falls to Mr. Khan. He had left the exhilarating task of reading aloud the long, long list of names of unknown clients and vanished victims to his young Khmer colleague, but now he's growing impatient. He slips a note to his associate, who tries to do better, or at least read faster, but there are simply too many names on the list, and reading them aloud is taking longer than anyone anticipated. Could we abridge it a little, perhaps? Could we skip the names of a few victims?

Mr. Khan passes a second slip of paper along to his young deputy. Then he interrupts, again, forgetting that honor once lost never returns. Mr. Khan begins by saying that no one on his side of the bar believes that Duch was without autonomy at S-21, or that he was just a cog in the machine, or that he had no way of reducing the suffering in the prison he ran. Mr. Khan speaks long enough to say that Duch's actions at S-21 patently heightened the regime's paranoia, and that they perpetuated a vicious cycle that led to even more people being arrested; Mr. Khan asserts that Duch did what he did not only because he wanted to join the ranks of the powers that be, but because it made his own life easier; Mr. Khan says that the truths to which Duch has admitted during the trial do not add up enough to dismiss the conviction that, when all is said and done, what we have sitting before us in the courtroom is an instance of "blatant denial."

A mobile phone rings. The presiding judge declares that Mr. Khan's time is up. Mr. Khan asks for five or ten more minutes. "You can have three," says the presiding judge—barely enough time for Mr. Khan to stop talking.

Some people find being exposed in a prestigious court as burning as acid on limestone. During the first weeks of the trial, Hong Kim Suon occasionally seemed inspired, and distinguished himself on the floor. Then, little by little, he seemed to deteriorate. After his more-virtual-than-real copilot, Mr. Sur, left him to fend for himself for six months of trial, Hong Kim Suon gradually foundered, losing the thread of his argument to the point that it sometimes seemed as though he had lost all sense of reason. On this last day, he is completely at sea. He spends half an hour speaking from the bar without bringing up a single point that the judges might be able to use in their judgment, and manages to tell the stories of only two of his ten clients. Mr. Sur, acutely aware of the limited time he has in which to present his argument, starts tapping his foot impatiently and directing appalled glances toward the corner of the public gallery where his team of handsome lawyers have gathered.

But Hong Kim Suon is intoxicated with the power of speech. To a lawyer, the time spent sitting and listening without distraction is substantial. A lawyer will spend hours passively waiting for an opportunity to present itself. Then he will spring into action and, preferably with some gumption, put forward an argument that is pertinent, effective, and eloquent. The challenge of finding the right words with which to break the silence is what makes lawyers so hungry for the opportunity to speak, and why they consider such opportunities so precious.

Hong Kim Suon has the floor and won't give it up; Mr. Sur can't take it any longer. The Frenchman gets to his feet and does something that reduces an awkward moment to a clownish incident: he pushes his colleague aside and grabs the microphone.

International courts have always been vulnerable to the incongruities, fantasies, and vagaries of post-colonialism. There's always the risk that the West—or the UN bureaucracy that its dominance sustains—will create situations in which the strongest, the richest, the most educated, and the most powerful will judge the weakest, the least educated, the most destitute, and the outcast. Many of those who work in international courts manage to conceal the awkwardness they feel by tempering their behavior and downplaying their own cultures.

Mr. Sur knows no such precautions. He is no more aware of the thunderous laughter taking place in the media room at his expense than of the vast chasm separating him from his Khmer audience. Mr. Sur's overcooked intonation, his feigned silences, his rehearsed pauses, the sheer, affected theatricality of his lawyerly posturing, and his inability to resist scattering a little stardust around ["I am Ingrid Betancourt's lawyer"] might come across as graceful in a Parisian court. But here in Cambodia, they just reinforce his bad reputation. And when Mr. Sur confuses Son Sen, Duch's boss at S-21, with Vorn Vet, who was executed there, or when he goes off on a tangent about the "thousands of photographs" that Duch supposedly had taken of himself at his home, he comes across as an amateur. What interest he still held as someone new and exotic evaporates. A sense of shame fills the courtroom.

Mr. Khan, meanwhile, nods off, wakes up, then nods off again. The cause is not the performance by his colleague, but rather jet lag. In a trial like this you can't get away with absenteeism and sloppiness, and everyone has been waiting for Mr. Khan and Mr. Sur to return from their truancy. Now that they're here, they're regarded with the sarcastic, contemptuous eye usually reserved for impostors.

# CHAPTER 36

IN HIS THICK TROUSERS AND TIGHT-FITTING, PALE YELLOW TUR-tleneck, Duch is dressed for winter. It's the first morning of the closing statements, and he makes a point of not looking in the direction of the lawyers representing the civil parties. He is meticulously annotating a document. Two months earlier, the hearings ended with a definitive separation between Duch and the victims. Now, while a Cambodian lawyer evokes the unfathomable pain, the crippling anxiety, the gut-wrenching cramps, the sleepless nights of the victims' families, Duch reads his document.

Perhaps it's the play of the light or the reflection of the glass pane, but Duch's face looks wan and sickly, like that of some half-human comic book character. By the afternoon, he finally deigns to look at his opponents, and even appears to be listening. Perhaps that's because the speaker is the only person on that side of the courtroom to have left the door open to Duch, if ever so slightly. "You have shed tears, that's the beginning of repentance; you have apologized, that is the beginning of understanding and of taking responsibility," says the lawyer representing the victims. "Now look at them, Duch! Look at these men and women you wanted to smash!"

Duch looks at the lawyer, who continues:

> You can't smash human beings, because one day they will rise up and come after you to settle scores. Maybe, just maybe, the victims will forgive you. But you cannot imagine how desperately these people are searching for answers. They want to know how one man, no worse

*than any other, can have behaved so barbarically. They want to know*
*how one ordinary person can be both respectable and terrifying at*
*the same time. François Bizot taught us that it would be a mistake to*
*think of you as nothing more than a coldhearted monster. We know*
*things are not that simple. But these people, these civil parties—whether*
*humble, poor, and uneducated, or cultured and well-off—are all*
*fighting the same battle. To be here. To tirelessly represent the rule of*
*law, just to keep going.*

For the first time that day, Duch is giving his full attention. The lawyer ends by reading the promise-filled preamble of Cambodia's Constitution, which was adopted a decade after the catastrophe ended:

> *We, the Khmer people, accustomed to having a grand civilization,*
> *a prosperous nation, a very large territory, a prestige glittering like a*
> *diamond;*
> *    Having fallen into a terrifying decay for the two last decades,*
> *when we have been undergoing unspeakable, demeaning sufferings*
> *and disasters of the most regrettable way;*
> *    In a burst of consciousness, rising up with a resolute determination*
> *in order to . . . guarantee human rights, to ensure the respect of law . . .*
> *we inscribe . . .\**

In the front row of the public gallery, next to monks clad in ochre and saffron, sits a group of Buddhist nuns wearing white robes. With their cropped gray hair and thick, Coke-bottle glasses with the dark-brown plastic frames that were once so popular, they look positively ancient. All these women lived through the revolutionary period of Democratic Kampuchea, when Buddhism and "reactionary religions"

---

\*  This is taken from the "unofficial translation of the Constitutional Council" of Cambodia in 2010 and corresponds to what was said by the civil-party lawyer in court.

were abolished and pagodas turned into prisons. Duch is no longer a Buddhist like his ancestors were, nor atheist like his former Communist idols, yet when he comes into the courtroom he duly acknowledges the monks and nuns, provoking sneers and ridicule among the civil parties.

If the victims of S-21 were still alive, says the Cambodian chief prosecutor, they would fill this amphitheater twenty-four times over.

"The regime was known as Democratic Kampuchea but there was nothing remotely 'democratic' about the three years, eight months, and twenty days in which the country was torn apart and more than 1.7 million of its citizens massacred," she says.

This trial, continues the prosecutor, isn't about the brutal, forced evacuation of the cities, that horrific rustication that alone left eighty thousand dead; nor is it about the forced labor camps, the so-called cooperatives; nor is it about the families torn apart, or the thousands who died of hunger, illness, and exhaustion. "Rather, this trial has focused on just one aspect of the regime: the enforcement of radical ideology that involved ruthless political violence."

The Cambodian prosecutor doesn't hesitate to link S-21 to Stalin's NKVD and Hitler's Gestapo. She links Communism to National Socialism, not dwelling on it but not abashed about it, either.

Next, it's the Australian prosecutor's turn to give a final statement. He is keen to make sure that Duch's role and the extent of his power at S-21 are clearly defined.

*Do you believe him when he says he was both a hostage and a prisoner of the regime from 1971 until the mid-1990s? A prisoner and a hostage forced to kill and torture human beings on a daily basis against his will and under the threat of death, with no choice and no chance of escape? Was the author of these crimes in reality a victim of the system? We have stated that the accused was neither a prisoner nor a hostage nor a victim. The evidence proves [it]. It clearly demonstrates that he was an idealist, a CPK revolutionary, a crusader who was prepared to sacrifice everything for his cause; prepared to torture and kill willingly for the good of the Revolution, no matter how grotesquely misguided it was.*

The prosecutor's fire soon fades and gives way to a colorless, professional, lawyerly tone. Yet even during the most monotonous moments, Duch never appears tired. Not once throughout the trial has his physical strength failed him. He still decides when to pay attention and to whom. He gave the last lawyers for the civil parties his full attention; but as soon as the prosecutor appears, Duch directs his attention elsewhere. He stares at the CCTV screen on his desk and slumps back in his chair, his whole posture one of disrespect. A moment later, he turns his gaze away from the speaker. It's the first time he's shown such glaring contempt. He's no longer the least bit interested in the prosecutor, and he shows it.

Of course, what the prosecutor has to say about Duch must be hard for him to take. The prosecutor makes it clear he doesn't believe a single word uttered by this master of confessions. He paints a portrait of a man completely devoted to Communism and more than willing to eliminate its enemies; a man who developed close personal relationships with the Party's most powerful leaders and who continued to work for them for some fifteen years after the fall of the regime; he describes a chief of police who helped the regime identify and destroy its enemies and who fueled the leadership's paranoia in the process. What's more, only Duch had the power to execute his own staff, and killing them was his decision. When it came to his own staff, Duch had options, says the prosecutor. He chose not to exercise them.

*We've all seen firsthand during this trial that the defendant is a meticulous man, a logical man bordering on the obsessive, a master of detail with a brilliant, albeit selective, memory. There is no doubt that under his authority, rules were always obeyed and order was always maintained. This is remarkable, given that the staff under his control at S-21 numbered more than two thousand.*

Throughout the course of the trial, the prosecution has often tried and failed to prove that Duch had the authority to order arrests. Now the prosecutor says that it doesn't matter if the evidence doesn't stack up. The important thing is that through the interrogation and torture he oversaw at S-21, the defendant caused hundreds and even thousands of people to be arrested—those famous traitors' networks. Duch was merciless, even with his own former teachers and friends. When his role became essentially to kill members of the Khmer Rouge, he was relentless. Duch's first revolutionary mentor and former teacher, Ke Kim Huot, was tortured, degraded, and executed at S-21, as was his wife. He was beaten, given electric shocks, and forced to eat spoonfuls of feces. She was raped with a stick.

"We submit that infliction of pain was not something he hated," the prosecutor continues.

> It was something he found both necessary and perversely gratifying. You have seen the evidence of the documents containing the accused's direct written orders to kill. They are chilling in their unemotional, unapologetic, ruthless efficiency. On a list of seventeen prisoners, including nine children, he simply wrote, "Uncle Peng, kill them all." On another he wrote, "Interrogate four; kill the rest." Sometimes he simply ticked off names with the annotation "Smash." Your Honors, when you review the evidence of the accused's efficiency, initiative, dedication, drive, enthusiasm, and zeal, there is absolutely no doubt that as misguided as he was, he was a fully willing participant in these crimes. The defense would like you to believe that the accused carried out his work while hating every moment and only acted in fear. All of the evidence in this case clearly disproves that hypothesis. The accused must accept reality: that unless he faces up to the truth and admits that he committed his crimes as a devoted follower, with the enthusiasm and zeal of an ardent revolutionary, he has not accepted full responsibility for his crimes before this court. We recognize that he's admitted the majority of the underlying crimes at S-21 and his responsibility as its director, and yet you must view his alleged

*remorse in the context of his continued refusal to admit his active and*
*enthusiastic participation in these crimes. He accepts responsibility*
*only on his own terms, wherein he attempts to paint a picture of*
*himself as an unwilling participant caught up in a machine he could*
*not escape, trapped by secrecy and terror. You must not allow him to*
*hide behind these false claims. You must recognize that he was not a*
*victim of the system but rather its loyal and dedicated agent.*

**THE PROSECUTION'S ARGUMENT,** based on documentary evidence, is irrefutable and convincing. In view of the notes Duch and others made at the time, it would require a huge mental leap, together with a degree of schizophrenia, to give serious consideration to the portrait painted by the defense. But the overwhelming truth contained in the S-21 archives isn't enough for the prosecutor. He asserts, but cannot prove, that the defendant regularly visited the main part of the prison, and specifically they interrogation rooms, even though most of them were in fact outside the prison itself. He claims that Duch participated directly in torture. As proof, he invokes Bizot's account, as described in his book, of Duch hitting prisoners "until he was out of breath," even though everyone, including the writer himself, has admitted that creative license doesn't amount to evidence.

In the end, people in the court begin to feel as though what happens during the trial doesn't matter; it's not important whether the case has been convincingly made or not. It seems the six months of legal arguments are irrelevant, since all sides are sticking to the positions they held at the outset, come hell or high water. It is against this background of indifference to detail that the prosecutor paints Duch's portrait, a portrait that is and must be harsh and brutal, but which nonetheless raises an uncomfortable question: does this urge to blacken a picture already darker than ink merely manifest our own irresistible yearning to distinguish ourselves from the monster at any cost?

In the fifteen years since international courts were first set up in The Hague, Arusha, Freetown, and Sarajevo, no prosecutor has given so condemnatory a closing statement against a defendant who has not only acknowledged the bulk of his crimes but also collaborated with the judicial authorities and provided damning testimony against other defendants. The international prosecutor is conscious of the special treatment he's giving Duch. He argues that this case is different from the rest. "Let us remember that, each time he has been backed into a corner concerning his own level of involvement, the defendant has been systemically unresponsive and, in our view, dishonest."

So Duch, unlike all other repentant perpetrators of crimes against humanity, is dishonest. That justifies the special treatment he receives, the logic goes. But the sad truth is, I don't know of a single "honest" confession any prosecutor has received in an international court. A confession is always the result of some compromise, some agreement, some deal; and in some cases such deals have been sufficiently opaque to mask a degree of dishonesty shared by all of the parties involved. Yet prosecutors from The Hague to Arusha have, without exception, proudly and assertively defended these "guilty pleas."

There is no such thing as an honest or dishonest confession. Whether it's obtained by the threat of a guilty verdict in a legal system that respects the rights of the accused or extorted by electric shock, a confession is always the result of an expedient settlement made between someone in a weakened mental state and mundane interests of the other side.

While the prosecutor pleads his case, Duch shows what he thinks of him. He gazes at the ceiling or away from the prosecution. His eyes are open and his features tired; his mouth is fixed in a grimace of either bitterness or irritation. The master of confessions has lost the game; the investigating judges have already spent a long time questioning him; he has given plenty of testimony against himself and against the four remaining Khmer Rouge leaders who are to be tried after him. The prosecutor has all the evidence he needs from Duch, at least on paper. Duch appears to have given away for free what few trump cards he had, without obtaining any guarantees in return. For someone so well-versed in power dynamics, it's a surprising mistake.

Duch looks right, then behind him to his left, then glances briefly at the prosecutor. Then he turns his back on him once again.

The Australian deputy prosecutor asks for a sentence of forty years. Usually a straightforward, affable, and accommodating man, he suddenly exudes an uncompromising authority wholly absent from his more good-natured side and embodies it with ease and foul-mouthed eloquence. The hesitancy and weakness he displayed during the cross-examination have gone. Given the opportunity to sum up his arguments unchallenged, he comes into his own.

A recess is called. Everybody stands. Duch stares at his accuser. Then he walks over to the glass separating the court from the public gallery and waves at a young man. He smiles and faces the crowd streaming past the platform at his feet. Nic Dunlop, the man who had "compromised everything" when he identified Duch in 1999, walks by. He blushes slightly. Duch watches his back as he walks away. The defendant disappears through a side door only to reappear a moment later. He begins pacing around his side of the courtroom with his hands in his pockets. He walks up to the glass again and contemplates the empty gallery. Two members of the civil parties take offense at his presence; one leads the other out of the room. Duch chats with a guard, returns to his seat, smiles. Then he stands again and goes up to his Cambodian lawyer, Kar Savuth, who is having his usual rest. Phung Ton's daughter and widow are leaning against the wall at the top of the public gallery. They never take their eyes off Duch. They never let him have the place to himself. François Roux comes into the gallery and sits with his family for a few moments. People start filing back into the room. Phung Ton's daughter and widow return to their seats, which are directly in front of Duch and Kar Savuth. Duch is smiling. He appears relaxed.

When the proceedings resume, it's Duch's turn to speak. It will be his final major statement. He talks ad nauseam about the endless killing perpetrated by the Communist Party of Kampuchea. He's convinced that he owes his survival, he says, to three things: never ordering an arrest himself or overriding his superiors' authority; never profiting materially from the war or the victory; and never conducting himself immorally with women. Obedient, selfless, and proper—he knew how to control his emotions. Duch meanders through a story of bloodshed and terror that is, by turns, absurd and tedious. He lingers on certain anecdotes and repeats, for the umpteenth time, theories or explanations that he has already given in court. His statement is muddled, meaningless, repetitive, and speculative; it involves a tale about the regime's top leaders killing each other off over an invisible and silent river overflowing with corpses. He finally concludes that the "Khmer Rouge regime wanted to use the killing to establish its dynasty in Cambodia and satisfy its ambition." As a Party member, Duch acknowledges his own responsibility for the last time. The way he puts it, however, has a deeper significance: "I clearly understand that any theory or ideology which mentions love for the people in a class-based concept and class struggle is definitely driving us into endless tragedy and misery."

This is no longer the story of a road paved with good intentions. Now Duch is saying that the worm was there right from the start, in the original philosophy. No doubt there are many former Communists and members of the Old Guard still clinging to the Revolution—including some who work for the tribunal and even for the office of the prosecution—who would struggle to articulate this criticism of the ideology they once served as lucidly as Duch does.

Sometimes, we get a glimpse of the old revolutionary Duch; sometimes, all the excitement of that era comes back to him, rising from the pit of his stomach and overwhelming him, if only briefly. These feverish outbreaks are striking. For a few seconds, the old Khmer Rouge soldier resurfaces, his faith apparently undiminished. The tremor that these reminiscences produce in court is proof enough, for some, that the apparatchik remains as committed to his cause as ever. Once a Khmer Rouge, always a Khmer Rouge, it seems. But behind the mask called Duch, Kaing Guek Eav is trying desperately, and despite everything, to exist.

> *I still maintain that a decision to choose which path to walk is made in a matter of seconds. However, the repercussions of making a wrong choice will result in lifelong remorse. Convinced that I was contributing to the liberation of the nation and its people and hoping that I would be serving my people, I devoted myself, my strength, my heart, my intelligence and everything else, including my readiness to sacrifice my own life for the nation and the people, to the cause. But I found I had ended up serving a criminal organization which destroyed its own people in an outrageous fashion. I could not withdraw from it. I was just like a cog in a machine. For the victims of S-21 and their families, I still claim that I am solely and individually liable for the loss of at least 12,380 lives. These people, before their deaths, had endured a great and prolonged suffering and countless inhumane conditions. I still and forever wish to most respectfully and humbly apologize to the dead souls. As for the families of the victims, my wish is that I will always maintain my humble and respectful behavior by asking you to kindly leave your door open for me to make my apologies. I promise I will do everything for my people, should they need me, in whatever circumstance in the future.*

For Duch, the intimate is still too perilous a field in which to tread. In the end, it's Duch the bureaucrat who prevails. He begins reading the thirty-four footnotes attached to his statement. Duch is now at the peak of his weirdness; incapable of expressing emotion and presenting a final statement which is the perfect illustration of his need to intellectualize everything, including his remorse and request for forgiveness. It should be a dramatic moment, yet it's quite banal. When Duch starts reading his footnotes, what little electricity there is in the courtroom immediately flickers out. We're left with Duch the archivist, deep in his papers.

"It's disconcerting," says one of the lawyers for the victims.

*Disconcerting because it leads one to think that this gentleman hasn't understood a thing; that he's still using his same method, including his famous footnotes. It's as though he's still living under the regime he chose to serve. He is still in the middle of the most absurd bureaucracy, one that crushes reflection, reason, and sensibility.*

When he's finished reading, Duch carefully tidies up his papers and slides them into a plastic folder, which he then hands to the court clerk. It's lunchtime. Tioulong Raingsy's sister leaves the court with tears in her eyes because the prosecutor didn't ask for life imprisonment. After the recess, her nerves still on edge, she wonders with trembling anxiety, "If the forty years are served in full"—she repeats the words, *served in full*—"then it's okay. But will they be?" Others seem to accept the prosecutor's decision to ask for forty years in prison. A phalanx of journalists bristling with cameras and microphones swarms after the survivor Chum Mey. Over the course of the trial, the former mechanic has turned into the victims' unofficial spokesperson. He has become astonishingly media-savvy, and it's both comforting and worrying to see his newfound skills exposed to all sorts of media, some maybe less well-intentioned than others.

# CHAPTER 37

**K**AR SAVUTH, WHO CLAIMS TO BE SEVENTY-SIX, HAS A SURprisingly young physique and takes good care of himself. He is nimble and blessed with a survival instinct forged by the several authoritarian regimes that have ruled modern Cambodia. He has the charisma and guile of one who is no longer afraid of anything, and a deft sense of theater. Though spry for his age, he has only so much energy, but he knows how to make the most of it. He lost family members under the Khmer Rouge, including two brothers, but he has been Duch's lawyer ever since Duch was arrested in 1999. Kar Savuth is also one of Prime Minister Hun Sen's legal advisors.

During an initial hearing before the international tribunal, a year and a half before the start of the trial, Kar Savuth sidestepped an uncomfortable question by flatly declaring that Duch had been tortured and badly beaten while being held in detention by Cambodia's military police between 1999 and 2007. It was a new and serious allegation, but when a judge asked him for clarification the following day, Kar Savuth quickly retracted it in a smug tone: "I didn't actually say that he was beaten up before the military tribunal."

Kar Savuth shares with many of his Cambodian colleagues a tendency to exaggerate, and one very quickly develops the habit of only half-listening to their over-embellished arguments. Bellicose and grandiloquent one moment, courteous and back-pedaling the next, Kar Savuth can charge through the legal arena like a bull bled alive; he can switch from hostile to obsequious in the blink of an eye. We have all learned not to take his bombast too seriously.

Kar Savuth gave one of his spectacular, impassioned speeches at the start of the trial.

"Why are we prosecuting Khmer Rouge leaders?" he asked.

> *There are three reasons: to see that justice is done on behalf of those who perished and those who survived; to prevent another such regime ever again surfacing in Cambodia; and to defend the nation's sovereignty. Who are the most important leaders of Democratic Kampuchea? How many of them were there? We cannot accept the legitimacy of this trial until these facts have been established. It is better to prosecute no one rather than to judge only some.*

At that early stage of the trial, everyone found the old fox's lively, flamboyant posturing in court, as well as his sheer nerve, entertaining. Everyone assumed that he would bluster on until he ran out of steam, then return to his charming and courteous ways. But Kar Savuth didn't run out of steam. He claimed that there were fourteen top leaders in Pol Pot's regime. Duch wasn't among them.

"If those fourteen people aren't prosecuted, it's a breach of the law! This trial must be stopped immediately!" he thundered before an audience that was both amused and embarrassed.

This time, the lawyer had gone too far. The prosecutor warned him and asked that

> *the chamber request the defense to clarify if these proceedings are now, at this point or at any other point, to be challenged on their legality? This chamber cannot proceed at this trial without having a clear answer from the defense as to its position on the legality of the prosecution of this individual. There is an expression that says "you cannot have your cake and eat it."*

The judges felt obliged to ask Kar Savuth to clarify his position. He immediately defused the bombshell with which he had just threatened the court.

"These arguments are just comments for the consideration of the chamber," he said with a disarming grin. "I'm not questioning the chamber's authority. I am quite aware that I could have raised this during the initial hearing if I had wished to do so. They are just my own comments."

Phew.

The defense floundered until François Roux returned to the helm and steered it toward the goal he had set for it a year and a half earlier. "There's no difference between international and national judges; there's no difference between international and national lawyers; and there's no difference between international and national prosecutors." Roux would like to say that the fool's game that he and his Cambodian partner found themselves caught up in is one being played at every level in the tribunal. Everyone might be sitting next to his Judas. Anyone might find himself betrayed at any moment. And it's hard to resist twisting the knife in your neighbor's wounds while waiting to be stabbed in the back yourself. Roux battered his opponent's already bleeding injuries.

The same day, the head of the Cambodian government declared that no more than five suspects would appear before this tribunal. Yet four months previously, after a year and a half of political wrangling, contortions, and delicate negotiations, the international prosecutor had requested that six other former Khmer Rouge leaders face prosecution. But his Cambodian colleague, all too familiar with the powers that be, had opposed these new indictments. Everyone in this tightly controlled court is vulnerable to such discord. Political betrayal is an old habit here.

Kar Savuth brought up the question of discontinuing the trial two or three more times over its course. But it no longer meant anything: in legal terms, it was too late. Everyone hoped that his repeated insinuations that Duch shouldn't be judged in this court were nothing more than his own hang-ups, a lawyer's itch to play the gadfly, or simply the result of the Khmer tendency to think in circles.

**THERE'S NOTHING STRAIGHTFORWARD ABOUT** defending Duch. Even those well-versed in defendants' rights, even those publicly committed to human rights, or who have made careers out of defending them, sometimes conflate the crime with the criminal and the criminal with the person defending him in court. Many human rights activists and lawyers have found such stigmatizations impossible to resist. And who hasn't wanted to write a letter to the editor such as this one, penned by a no-doubt well-educated reader of *Le Monde*, a highly respected French newspaper:

> I hope that the defense lawyers had no family, friends, or acquaintances among the 14,000 people that this vile individual ordered put to death. Had Cambodia kept the death penalty, he would deserve it twice over. What a cynical move by the old man. In any case, he won't enjoy life for much longer and if we only let the people carry out their own justice, he would be stoned to death on the spot. And to think that there are lawyers who accept the vile job of defending him!

Behind this brave vigilante's words we can hear the lynch mob's call; we see its redemptive fist smash into Savuth's nose and Roux's temple, the throng frothing at the mouth; we hear the hollow sound of stones piling up on the bastard's corpse while nearby in a corner, the lawyer pisses vile blood onto his worthless robe.

We need only watch how men and women of all social classes react to the trials of torturers and executioners to gauge our collective lust for the gallows, for the firing squad lurking in a football stadium, or for watching women with the shaved heads in 1945 at the French Liberation. In every trial I've covered—whether in Africa, Europe, or Asia—I've felt the breath of that bloodlust, the hatred that exists even among the most well-educated, on the back of my neck. It is the same as the rush of air that preceded the blow of the pickax handle that was the last thing felt by the victims at Choeung Ek, on their knees at the edge of the pit.

Kar Savuth knows all of this. On the day he makes his closing statement, he begins by asking his countrymen for their understanding for the "vile job" lawyers are obliged to undertake when defending a man like Duch. With that caveat out of the way, he returns to his favorite theme: there were hundreds of prisons, some of them even more murderous than S-21, yet their former directors remain free, untroubled by the law. Why is Duch and Duch alone standing trial? Duch is just a scapegoat, claims the lawyer, and scapegoating isn't justice. Nine months after the trial began, Kar Savuth is sticking to his guns: this tribunal has the authority to judge the top leaders and main perpetrators responsible for the atrocities; Duch wasn't among those who decided whom to arrest and execute; therefore, he wasn't one of the fourteen people in charge of the Khmer Rouge. Of the individuals named by Kar Savuth, only three are still alive. All three between seventy-nine and eighty-five years old, they are scheduled to be tried after Duch. It's unfortunate, says Kar Savuth, but the court can't have it both ways: either charge all the former directors of all the other prisons in Democratic Kampuchea, or else free Duch.

Kar Savuth can show a good deal of finesse when it suits him. But today is one of those days that call for impassioned argument, which he gladly provides. The temperature rises, like a fevered trance taking hold of dancers swaying to primordial drumbeats in the sweltering night. Kar Savuth, intoxicated by his own words, grows heated and provocative; he finds it appropriate to invest Pol Pot's murder of Son Sen with some gravity; he deplores the Spartan conditions in which Brother Number One spent the last months of his life. Those crimes ought to have been punished, he says. The notion is so noxious that it is immediately dismissed by everyone present. But Kar Savuth has a knack for jumbling the coarsely preposterous with sensitive truths. Amid all his nonsense, he reminds us that high-ranking Khmer Rouge cadres are still serving in the army today, some of them as generals, and none of them are fearful of being arrested. So, really, why Duch and not them? The Cambodian lawyer has been preparing for his moment for a long time, and now, at last, he takes the final step: he declares that Duch is "not guilty" and that all charges against him should be dropped. Nobody finds this funny anymore. The atmosphere in the courtroom is electrified.

François Roux has been cut loose by the prosecutor, betrayed by his own cocounsel, and has helplessly watched his client withdraw into himself, all in a single twenty-four-hour period. The job to which he has devoted himself for the past two years now lies in ruins. Roux was Duch's best chance, yet Duch ditched him. He was the best ally the civil parties had, yet it's at his feet that they have laid all their anger and bitterness. He was the tribunal's best asset, since he promised it a dignified, unobstructed trial, yet he now stands humiliated before it. A good many people—even among those who consider Roux's a "vile job"—recognize that he has carried this trial, that he has stamped it with high standards, credibility, dignity, and competence. Yet two days before its end, he suddenly finds himself its biggest loser. He had wished for this trial to be transcendent; instead, he has fallen into the abyss. How can Roux fight the prosecutor when his own client and cocounsel behave as if they are on the prosecution's side?

Kar Savuth's *coup de théâtre* adds more drama to a trial that already has plenty; the latest one centers on a lawyer with a broken dream, who realizes that nobody else in the courtroom wants what he wants, a fact driven home during the trial's Shakespearean final act: a dagger in the back.

The victims' families, meanwhile, close ranks. For thirty years they've been trampled upon; now, finally, they have found strength in numbers, and they sit together in three long rows, persuaded that the drama unfolding before their eyes is proof of what they have long claimed: that Duch is a master manipulator, that his confessions are nothing more than a smokescreen, and that now, at last, the torturer is showing his true face. The Cambodian lawyer's outrage becomes their bitter triumph. It validates their desire to see Duch severely punished. How can they support François Roux when both his client and his cocounsel are fighting him?

# CHAPTER 38

"I'VE COME TO SEE *LE MAÎTRE,*"—THE MASTER—A VISITING AMER-
ican lawyer says, referring to Maître Roux, taking her seat in a
packed courtroom the following morning. The man she has come to
see enters the courtroom like a Roman warrior sacrificed to a *nauma
chia.*

François Roux might well say, *"Ave, populus, morituri te salutant"*
("Hail, people, those who are about to die salute you"). He is about to
make the last closing argument of his prolific career. He finds himself in the most indefensible position imaginable, caught between
surrender and humiliation, two sentiments he cannot stand. Yet even
dethroned, even beaten and humiliated, "the master" still stands lustrous atop some invisible pedestal. Even in the hour of his defeat, he
is still the one most admired by the public.

The master stands to speak, very slowly straightening his body.
His hand reaches for the microphone as though in slow motion; he
leans forward halfway at a right angle, as though his seat has remained
stuck to him, and then stays in that position, his torso barely moving.
His movements are so drawn out that once he completes them, you're
not quite sure whether he's standing entirely straight or not. Then his
confident and patient voice booms through the courtroom: "To stand
up and speak in someone's defense, that is what makes our job noble.
To stand beside the accused, beside the person here accused of one of
the most serious crimes imaginable—a crime against humanity. Imagine that: a crime against humanity."

Roux speaks in a natural tone that makes him seem close; his timbre is clear, his words carefully chosen. With just a few sentences, this intellectual tightrope walker succeeds in establishing a calm solemnity in the courtroom. Now he brushes aside yesterday's insult by turning it into a challenge for himself:

> *I have explained at length to the team that these are two contradictory things: we cannot, on the one hand, ask for the acquittal of the accused, which would mean that he is not guilty, as well as enter a guilty plea. It has now been publicly expressed that the accused will plead not guilty. I therefore withdraw the guilty plea.*

Once he has brushed his own team's betrayal under the carpet, he can continue to plead, despite everyone, on behalf of the Duch he has defended for the past two years, the one who repented, the one who broke down when confronted with the reconstruction of the crime at S-21 or at the mention of Phung Ton's name. "Who can dispute those moments that we have all lived through? Who is in a position to contest this? Didn't we all experience those same moments of utmost sincerity? Weren't we all utterly moved?"

The regime's leaders didn't need Duch to become paranoid, he says, quoting the much-respected historian David Chandler. "Yes, the paranoia began in the center of the Party, not at S-21. The paranoia began in the center and spread to the ranks. If the most dangerous enemy was the invisible one, then there could be no end to the terror, because the enemy could not be seen."

Roux must also nullify the damning portrait of Duch painted by the prosecution, of a man who committed his crimes with particular enthusiasm. The master refers to Article 5 of the Party statutes, which each member had to scrupulously obey:

> *I quote: "One has to exercise initiative as well as autonomous creativity, a dynamic work ethic and consistent, intensive work methods." That means that it was incumbent upon each and every*

*Party member to display enthusiasm, the very same enthusiasm which
the prosecutors now hold against Duch. Come on! You have seen,
as I have seen, the propaganda films showing all the people building
dams! Have you seen them crying or have you seen them singing? And
you want to blame Duch for doing something that was demanded of
everyone?*

Duch said what needed to be said, argues Roux, even if he didn't
say everything.

> *Ms. Studzinsky tried to learn more. We were unable to do so. Is
> it that this trial has turned into one of evasion? I guess so. When we
> spoke of Phung Ton, you saw the emotions it triggered in Mam Nai
> as well as in Duch. What's at stake here? I don't know any more than
> you do. I am only describing what I see. So, yes, there's no doubt that
> gray areas remain. No doubt there remain things that are hard to
> admit. But so it is.*

IN HIS GRIPPING PLAY *INCENDIES,* one of Wajdi Mouawad's characters
hides his criminal past from his family. The playwright gives his character this line of terrible, simple veracity: "Why didn't I tell you? Some
truths can only be revealed provided they have been discovered."

At this stage, François Roux appears to be succeeding in his effort to make the court forget about Kar Savuth. A good lawyer knows
how to spin a story without appearing to; he knows how to weave,
from scattered, flimsy scraps of information, a thread that appears as
strong as one of those climbing ropes with which you can confidently
rappel down a cliff. The master deftly and boldly strings together
documents and quotes to forge a harmoniously structured argument,
rich in substance.

Duch never takes his eyes off him. He removes his interpreter's
headset, leans forward, and listens closely in that language spoken
by so many Party intellectuals: French. The lawyer raises his voice to

show his indignation at the proposition that Duch had access to the secrets kept by the Party's highest-ranked leaders; his voice is strident when he defies anyone to claim that one could disobey the Big Brothers; then, as he explains that there was no loophole, no way out of the endless cycle of death decreed from on high, his tone drops to one of great weariness.

The master mentions his own broken dreams. He tells of his dashed hope that the office of the prosecution would pursue a dialogue with the only major Khmer Rouge cadre to have accepted responsibility for his crime. Roux had dreamed of a prosecutor who would rather outdo himself than give in to public opinion. He denounces the "conventional, traditional argument whose underlying philosophy is 'This man is a monster.'"

> *Members of the prosecution, yesterday you vehemently spoke out against a man who was on his knees begging for forgiveness. I do not like this. I do not like the tone of your statements or of your arguments. A scapegoat is one on whose head all the evils, all the sufferings of a society are loaded. Among the Hebrews, the goat was sent into the desert so that society could redeem itself by saying, "This goat bears all our wrongdoings." That is a scapegoat. Duch does not have to bear all the horrors of the Cambodian tragedy. No, Duch is not the person described to you by the prosecutors.*

During a pause in proceedings, Nic Dunlop turns to me with a mischievous look in his eye and says, "I've never heard such an eloquent 'fuck you' before."

The master continues his argument. He talks about the crime of obedience, and its twin, the crime of submission. Roux had naively hoped that he and the prosecution would steer the trial toward the same destination, the one that François Bizot set forth during the opening of the trial and which David Chandler defined in the conclusion of his work: namely, that to understand S-21, we needn't look further than ourselves.

*Are we just going to rehash the same old arguments before this court? "He committed crimes, so therefore he must be convicted and society will be better for it." Or even worse: "This will not happen again." Well, let me tell you: it will happen again, as long as we avoid looking with clear eyes at the phenomenon that turns a normal man into a torturer and executioner. This phenomenon, this crime of obedience, is exactly the subject that Chandler so courageously tackled.*

*For thirty-five years, I have been hearing about this experiment, this terrible experiment where they took ordinary American citizens just like you and me, put them in a room, and said, "You see, behind the glass pane, the person sitting there in that chair? He's lying. He has electrodes on him and you have a button. Each time you ask him a question and he lies, send him a jolt." There's an instructor wearing a white coat next to you, telling you, "It's okay, go ahead." Sixty percent of the people—people just like you and me—obeyed the person in the white coat and kept pressing the button until they reached the lethal dose. The person on the other side of the glass pane was only an actor playing the role of the victim. What a terrifying experiment. I have heard people here in this room tell the court, "I can't answer that, I have to go and check with my superior." Yes indeed, that is how we all function.*

**THE END OF A TRIAL** is like the end of captivity. What has taken months—years, even—all those depositions, testimonies, facts, documents, experts, and pleas—evaporates and is forgotten behind the one and only thing that everyone will remember, and which erases everything else: the verdict.

"So, we come to the sentence," says the master quietly, in a silence so intense that when he rustles the page in front of him, it sounds like a judas hole being slammed shut.

Previously, Ouk Ket's widow had suggested, with bitter sarcasm, that Duch should be condemned to working as a groundskeeper on the killing fields of Choeung Ek, which, she indignantly pointed out, were haphazardly maintained at best. The master, a little carried away, now allows himself to make a similarly coarse statement. Like a climber carefully testing where he places his foot, he suggests that Duch ought to be condemned to giving guided tours of the scenes of his crimes, to show "younger generations what should not be done." To plead in court means to sometimes risk going too far.

It is time to reach a conclusion.

> *It's always a difficult moment for a lawyer because he knows he can say no more after this, so he asks himself, "Have I said everything that needed to be said? Have I done enough?" Duch, all your victims were your brothers and sisters in humanity. You said that you were a coward and that you did not go to see them while they were in detention. In the eyes of humanity, you will never be absolved of these crimes, and the gazes of those you did not wish to meet will follow you forever. But what about us, Your Honors? Are we prepared to look Duch in the eye and see him as our fellow human? And will you, by your decision, bring Duch back to humanity? For quite some time now, lawyers across the world have argued that one of the goals of sentencing is to rehabilitate the guilty party. But is rehabilitation forbidden in the context of a crime against humanity?*
>
> *A final word: Duch is dead. Today, his name is Kaing Guek Eav.*

And with that, Roux takes his seat as slowly as he stood a few hours earlier before the curious crowd; he moves as though he were in the twilight years of life.

# CHAPTER 39

**E**LOQUENCE IS LIKE A BUTTERFLY—ITS LIFE SPAN IS EPHEM-
eral. Now that Roux is back in his seat, the lawyers for the
victims and the prosecutors are determined not to let things stand as
they are. For the past twenty-four hours, they've watched, goggle-eyed,
as the defense team imploded. Now it's time to sound the horn for the
final assault. Mr. Sur has already returned to France, but Mr. Khan
seizes the moment to salvage his reputation. He is the first to pounce
on the opportunity presented by the unbelievable discord among his
opponents:

> *The defense is putting forward two completely different positions.
> This is wrong. It is unfair to the people of Cambodia, it is unfair to the
> victims, and it is unfair to Your Honors' search for the truth. Instead
> of responding to the views, the concerns, the pain, and the plight of
> the civil parties, the accused embarked upon what I call a carefully
> constructed, paragraph-by-paragraph, footnote-by-footnote statement.
> Thirty years after the events, our civil parties want the truth, they
> want their lives back, they want some kind of closure, and it is simply
> unacceptable that we are left in this chaotic state of affairs where we do
> not know actually what has been said. Mr. Roux states that the accused
> is not pleading guilty but he is accepting contrition. His Cambodian
> lawyer says, "Release him, he's absolutely not guilty." This is a unique
> case that can be described by many adjectives, but I'll simply say [that
> it is] highly unacceptable and absolutely avoidable.*

Silke Studzinsky sees the defense team's implosion as a slap in the face to the victims:

> After the accused and his defense tried to convince the civil parties that his partial admission amounts to a truthful, sincere, and genuine confession, the civil parties now became even more convinced beyond any doubt that the accused was and is playing a game, and that at the very least the time has come to shed the sheep's clothing. The civil parties who are seeking justice and truth are further alienated and more offended as the credibility of the accused is once again more corroded. His wish and demand to return to Cambodian society must be rejected.

The lawyer Hong Kim Suon, another of the victims' lawyers, begins by saying that he's deeply saddened by the defendant's about-face. Within a few seconds, however, he's in tears, quite overwhelmed by the pressure of the event. Throughout the trial, Duch had accorded this lawyer a special, almost fraternal, form of respect. Now he gives him no more attention than he does the others. At this point, Duch has started breaking away from the society to which he had hoped to be reinstated. In one of the interviews he gave to filmmaker Rithy Panh a few months prior to the start of the trial, he had a terrible intuition: "The day I am forgiven, I will prostrate myself in gratitude. If I am not forgiven, well then, let's leave it at that and wait for life to end." As his trial reaches its conclusion, Duch realizes that he is not going to get the forgiveness he had hoped for, and so withdraws into himself. He turns away from others. The previous evening, after Kar Savuth's bombshell, Roux had naturally asked to see the defendant. But Duch sent word through the prison warden that he was too tired. Just as he once avoided visiting his friends about to be killed at S-21, Duch now avoids a meeting that would bring him face-to-face with yet another betrayal.

**THOSE ON THE OFFENSIVE** are always more brazen than those in retreat. The prosecutor doesn't miss this undreamed-of occasion to rip into an opponent who has dominated him from start to finish over the course of the trial. He also uses the opportunity to justify his own intransigence.

"What has happened in this trial is that the prosecution, the civil parties have been grossly misled by the defense," proclaims the Australian prosecutor.

> *They have been saying throughout this case that certainly they would not be asking for an acquittal; and that's what they've done yesterday. That's loud and clear. Now, if that's the case, he should get no—no—mitigating factors in relation to his sentence, none at all, because that's not cooperation at all. So that's the assumption that we make, but I have a feeling that that's not the case. I have a feeling that counsel have acted without instructions and I think this needs to be resolved before we leave the courtroom. It's very, very unclear what the defense are, in fact, doing, but one thing that is clear is that from that defense yesterday, they asked for an acquittal of this accused, and I strongly suggest to Your Honors that you speak directly to the accused and find out whether they were acting on [his] instructions or not.*

Duch leans over his desk to confer with Kar Savuth. The reply to the prosecutor's challenge seems to get the Cambodian lawyer revved up. He nods his head very firmly several times, and slices through the air above his desk with both hands.

The defense's disintegration becomes blatant during the recess. Duch confers with Kar Savuth while the team's four international lawyers deliberate among themselves. Then Kar Savuth leaves the room and Duch sits alone. When Kar Savuth returns, Duch exits the room, as do Roux and his team; Kar Savuth remains alone in the courtroom. When Duch returns, he greets several former students through the glass pane and then returns to his box. François Roux reappears, puts his robe back on, and comes over to talk to Duch. Behind him, Kar Savuth rehearses, with much gesticulation, the answer he's about to give. A minute before the proceedings resume, the two lawyers exchange a few strong words. Kar Savuth quickly discusses something with Duch. The judges return to the courtroom. Presiding Judge Nil Nonn remarks that the statements made by the defense lawyers have been particularly incoherent. He then lets the Cambodian lawyer speak.

Either there's a problem with the interpreters or he is particularly muddled, but Kar Savuth's words come across as mostly incomprehensible. Still, though his reasoning is flawed, his conclusion is crystal-clear: Duch cannot be judged according to national law.

The old lawyer certainly doesn't find himself at a loss for words. In the middle of his diatribe, he mentions a few unsettling facts: for example, that three-quarters of the victims of S-21 were themselves servants of the regime and that, consequently, the tribunal ought not to make it its priority to give justice to all these Khmer Rouge cadres with blood on their own hands. But his speech is nothing more than a jumble of vague, incomplete, rehashed, or irrelevant ideas. After having confused the sentencing, the non-retroactivity of the law, and the reality of the state of war with Vietnam, Kar Savuth takes off his spectacles and, while the crowd murmurs, concludes: "Duch has been held for over ten years. Other wardens aren't in prison. Therefore, I think the time has come for the court to release my client and to allow him to go home."

When François Roux stands and straightens his frail body with his characteristic slowness, it's usually an internalized way of warming up for the coming challenge. This time, it's more like the movement of the condemned man laboriously carrying his cross up Calvary Hill.

Lawyers are sometimes called upon to sacrifice themselves. Roux uses all his rhetorical talent one last time in an effort to save Duch's skin, even though Duch has devoted himself to other saints. With so many adversaries in the courtroom, including in the seat next to him, Roux has the luxury of being able to choose among them. And once again he manages, by some subtle rhetoric, to take his ineffable teammate's arguments and to make them seem to fit naturally with his own. The man knows his stuff. But the prosecutor won't be fooled, and he sticks to his agenda:

> *The defense has avoided your [the judge's] question in relation to why [they have entered] this change of plea. On the one hand, we have the defense asking to "mitigate his sentence," and on the other hand, they're saying "acquit him." I think it's very important to find out why they are running these two defenses at the moment.*

The clock is ticking and Roux must try to beat it. He attempts a death blow:

> *I am sorry if counsel for the prosecution failed to listen to us closely enough. The word "acquittal" was never uttered this morning. Both counselors for the defense asked that the defendant's sentence be reduced, and that he be released as soon as possible, given that he has already served a ten-year term and has fully accepted responsibility for the crimes committed at S-21. Nothing has changed. It is not an acquittal. If this isn't clear enough for my learned friend, then I am sorry.*

But it isn't clear enough for Judge Nil Nonn, either. And he wants an answer. Duch finds himself ordered to speak. After having blatantly snubbed the office of the prosecution since the beginning of the week, he now finds himself looking directly at them. He gesticulates a lot, his movements quick and sharp. His Revolutionary French is a little rusty—"democratic centralism" becomes "centralized democracy"—but he holds firm. He reminds the court that he has been cooperating with the law for the past ten years without fail. He emphasizes that he has agreed to acknowledge and discuss the crimes committed at M-13 even though they fall outside the tribunal's jurisdiction, and that he also agreed to talk about events that happened after 1979, even though the tribunal isn't mandated to cover that period, either. He repeats that he has accepted a broad and general responsibility as a member of the Communist Party for the totality of the crimes committed under Pol Pot's regime, and that he made an apology for those also, even though his case and his crime before this court are limited to what happened at S-21. Finally, he repeats his Cambodian lawyer's argument: he wasn't one of the regime's top leaders, and this tribunal was created to judge them and them alone. "I have been detained since May 8, 1999. It has been ten years, six months, and eighteen days already. Therefore I ask the court to release me."

The judges deliberate quickly among themselves. Then Nil Nonn asks Duch to stand once again. "The question is: are you asking the chamber to acquit you of all charges, or are you asking the tribunal to reduce your sentence based on your cooperation with the court and the time you have already served?"

Duch has his back against the wall. But he doesn't want to make a run for it, at least not alone. He says that his analytical skills are too limited, that he would like to be released but that he prefers to entrust himself to his lawyer. Specifically, his Cambodian lawyer.

Kar Savuth repeats his argument about the top Khmer Rouge leaders. Judge Lavergne puts his head in his hands. The lawyer tries one last trick: he simply doesn't answer the question.

Enough is enough. Someone is going to have to talk. Judge Cartwright takes the reins. "Counsel Kar Savuth, do I infer from your last comments that the accused is seeking an acquittal?"

Now it's Kar Savuth who is backed into a corner. Usually when he's caught in such situations, he beats a wily retreat. The entire courtroom is waiting with bated breath for his answer, on which the trial will end. "That's what I said. To release means to acquit."

Kar Savuth doesn't retreat. For the second time in two days, François Roux has been backstabbed by the man who has been his brother-in-arms for two years. But Cambodia has never been a merciful place.


**FRANÇOIS ROUX OFTEN LIKES** to say that to defend someone is to suffer alongside them. His client may have disavowed him, but Duch hasn't asked for his resignation or asked him to do anything professionally unethical. Being Duch's lawyer might not be a vile job, but it can be an incredibly lonely one. François Roux decides to stay in his seat while awaiting the verdict. To him, that's the best way to stay true to his oath to protect the defendant's rights—even if it is at his own expense.

The psychology of a prisoner is not the same as that of a criminal. A prisoner has different impulses and subscribes to different survival strategies than a criminal. Many people think that Duch's about-face is a sign of the old Khmer Rouge cadre rising from the ashes. But the about-face could just as easily be the consequence of the state of mind in which any prisoner facing a long sentence might find himself. Most prisoners are like Icarus: the mirage of freedom taunts them like a blinding sun. Duch's choice reveals to us a lot about him even as it confounds us. Some see it as the quintessence of either the master manipulator or the coward. Others speculate about some underhanded political negotiation, with Kar Savuth acting as a go-between. And others still see it as an illustration of the former revolutionary's inability to live by an ethic of responsibility, despite his aspiration and prior agreement to do so. Those who like certainty can have it. Those who tolerate doubt can keep tolerating it. Duch comes from a background in which survival trumps everything else, and he is showing us that it is the only thing that he strives for.

Wisely, the judges decide not to attach too much importance to it.

A month before the verdict is handed down, Duch sends a letter to the court's administrators, letting them know that he is dismissing François Roux. He never did like meeting face-to-face those he was betraying.

# CHAPTER 40

**D**UCH SAYS THAT AS A TEENAGER HE WAS DRAWN TO STOICISM. Traditionally, Cambodian children are taught very young not to complain. To complain is to reveal one's weaknesses, says one of the psychiatrists acting as an expert witness. In this cultural context, Stoicism, which urges its adherents to control and repress their emotions, finds fertile ground. Staying silent shows that you aren't a coward and that you don't give up.

For generations, every French school student has studied that poetic pearl, Alfred de Vigny's "Death of the Wolf." French colonial schoolteachers diligently imported the famous nineteenth-century poem into Cambodia. Its lines made an impression on a young Kaing Guek Eav. He says that when he was head of the M-13 camp, he used to recite to himself the poem's closing verses:

> *Moaning, weeping, praying is equally cowardly.*
> *Staunchly carry out your long and heavy task*
> *In the path to which Fate saw fit to call you,*
> *Then, later, as I do, suffer and die in silence.*[*]

Duch says the poem helped strengthen his resolve to carry out the cold and heavy task which the Party saw fit to give him. What can be more disturbing than a torturer reciting a great and noble poem?

"Sometimes," says Duch, "we must do a job we don't like."

---

[*] Bonhomme, Denise, *The Poetic Enigma of Alfred de Vigny: The Rosetta Stone of Esoteric Literature* (Victoria, Canada: Trafford Publishing, 2003).

Roux sets the stage well that day. After Duch recites his four lines of verse in French, the lawyer allows a heavy silence to fill the room. Then, amplifying it, he sits down without a word.

For the non-Stoic, the feeling it produces is spectacular. For the victims, it is excruciating. For the brother of Chum Narith and Chum Sinareth, that his brothers' executioner should invoke a poem or indulge in any other spiritual refinement is unacceptable: "After the defendant was finished, his lawyer let two or three minutes of silence pass," he angrily tells the court.

> *You couldn't even hear a mosquito buzzing, and you felt sorry for the accused. It was a clever technique. But if the defendant is comparing himself to the wolf, then he's an impostor! What bravery are we talking about? He knew that the professor was being tortured and dehumanized. What courage is there in that?*

Everyone wanted to hold the poem close, like some precious fabric found by heirs among a late relative's belongings. People fought over the death of the wolf.

"Do you realize that you have retained the most dangerous, the most lethal elements of the poem that leave no room for man?" says an affronted lawyer for the victims.

> *You have taken only those parts that fit your view of life, that is to say, that man is a wolf to other men. Do you realize that your response to being tried for crimes against humanity is to recite romantic poems? We are not here, sir, for cultural edification! This is not a literary salon! I am talking about twelve thousand dead at S-21! Some say sixteen thousand! What is romantic about that? We believe that you have become a wolf to men. Yet we do not wish for the death of the wolf.*

"**I AM NINETY-ONE YEARS** and six months old."

People in the twilight of their lives share with those at the dawn of theirs the habit of measuring their age exactly. For both groups, months matter. The very young and the very old keenly feel the value of every season. Each month is a palpable measure of how little they have lived or what little life remains to them. When the old man cheerfully and precisely gives his age to the court, a loud murmur spreads through the public gallery. He is the trial's final witness. People are asking themselves what he's doing there. What can this former French Resistance fighter, who was deported to Nazi concentration camps and who later became a diplomat and a fervent and unfailingly courteous defender of multiple humanist causes, possibly have to say in this place and on this matter?

It's an irrelevant question that doesn't occur to the Cambodians gathered in the public gallery. They respect their elders here. Simply being old is enough to generate respect. This venerable old man is testifying from Paris. The satellite link is mediocre at best, but his voice is sparkling and melodious. Those who have experienced mankind's folly as well as his genius can become either misanthropes withdrawn from the world, go dance with the real wolves, or aspire to acquire the wisdom of worthier souls. At ninety-one years and six months, Stéphane Hessel* has entered that curious kingdom from where so few of his peers can speak their minds. He seems able to express and understand everything with objective clarity. The patriarch is firm yet careful never to hurt anyone. He gets straight to the point without omitting anyone, particularly the weak and downtrodden. He knows that, despite our desire for easy conclusions, the world isn't black-and-white, governed by good and evil, or divided between torturers and victims. And he manages, without seeming to, to encourage others to think things through a little more deeply.

"Can this trial reach a conclusion that isn't unilateral?" he asks with a subtle and cryptic use of rhetorical questions and negations that cancel each other out. Hessel speaks a rich and lucid language rarely heard these days, one that rejects the excessive and superflu-

---

\* Stéphane Hessel died on February 26, 2013, at the age of ninety-five.

ous. The exact opposite, in other words, to the language spoken by certain lawyers. His clear, wide eyes and generous smile lend his kindness a quality that the foolhardy might mistake for naiveté. But Hessel is as immune to vanity as he is to the self-serving appeals of policymakers. He limits himself to making few statements and making them cautiously. But those he does make carry the moral weight of a man who, in the words of another poet and Resistance member, René Char, passes in front of the "snake-men" without crushing them.

Stéphane Hessel repeated poems to himself while being tortured by the Gestapo; Duch recited them to himself while torturing others. Roux knows that the venerable survivor can recite many verses from memory, and that poetry helped him to survive the death camps. He also knows that Hessel included Alfred de Vigny's poem in an anthology he published. The time has come for us to ask him how we, snake-men, should understand the poem.

The lawyer begins to read: "Alas! I thought, in spite of that great name of Men . . ."

Just then, we hear the satellite-borne voice of the old poetry-lover resonate from the far side of the world:

> . . . How ashamed I am of us, weak ones that we are!
> How one must leave life and all its woes,
> You are the ones who know it, sublime animals!
> Seeing what one was on earth and what one leaves,
> Alone silence is great; all the rest is weakness.
> Ah! I understood you well, wild voyager,
> And your last glance went straight to my heart!
> It said: if you can, have your soul arrive,
> By dint of remaining studious and thoughtful,
> All the way to that high degree of Stoic pride
> To which, born in the forests, I immediately rose.
> Moaning, weeping, praying is equally cowardly.

*Staunchly carry out your long and heavy task*
*In the path to which Fate saw fit to call you,*
*Then, later, as I do, suffer and die in silence.*\*

"These are the verses recited by the man who would become a torturer and executioner," says Roux. "What is this poem about Stoicism trying to tell us?"

"Naturally, I am always moved by a beautiful poem, but a poem merely reflects the position that the poet thinks men of honor ought to adopt toward even the most cruel vagaries of life," says Hessel.

> *If the defendant agrees with the text, he will have to endure his eventual sentence with the same strength, the same courage as the wolf that, a few stanzas earlier, "seized, in his burning maw/ the quivering throat of the boldest dog/ and did not release his iron jaws." I don't remember the lines exactly, but you see my point. If the defendant were to be absolved of the responsibilities he has taken upon himself, it would be counter to the choice he has made.*

"Is it possible for a man to redeem himself? Do you believe that redemption is possible?"

> *That is a difficult question and an embarrassing one when we think about the suffering of the victims, whose imaginations will forever be haunted by the memories of the terrible things done by the defendant. I have no doubt that the defendant will draw some benefit from everything that was discovered and said about him by those around him. That being said, I'm not sure that a truly honorable man could wish for anything other than fair retribution for the crimes of which he knows he is guilty.*

---

\*  Bonhomme, Denise, *The Poetic Enigma of Alfred de Vigny: The Rosetta Stone of Esoteric Literature* (Victoria, Canada: Trafford Publishing, 2003).

In just a few words, Hessel has settled the quarrel about the poem and restored poetry to mankind. Duch has risen from his seat to salute a witness only twice: once for Phung Ton's widow and once for his pastor. When the presiding judge indicates that the old man's testimony is over, Duch rises for the third time and does the *sampeah*, pressing his hands together in front of his face, slightly higher than Christians do when praying. Hessel watches him through the satellite link. Quite naturally and without a moment's hesitation, he puts his own hands together in front of his face and reciprocates the defendant's gesture of respect. There is no one that he would not acknowledge.

On July 26, 2010, Duch was sentenced to thirty years in prison. He immediately appealed his sentence. On February 3, 2012, the chamber of appeals gave him a life sentence. Poetry, alas, enlightens only free men.

# HISTORICAL MILESTONES

## 1953-1970: THE SIHANOUK ERA

On November 9, 1953, Cambodia gained its independence after ninety years of French colonial rule. In 1955, Norodom Sihanouk, crowned king of Cambodia in 1941 at age eighteen, abdicated the throne to his father in order to found his own political party. Prince Sihanouk won the elections and became prime minister. Five years later, he claimed the title "head of state." In 1964, he coined the expression "Khmer Rouge" to designate Cambodia's Communists, whose party, created in 1960, operated underground. The Khmer Rouge undertook their first armed operation in January 1968.

## 1970-1975: LON NOL'S REGIME AND THE CIVIL WAR

On March 18, 1970, Sihanouk was deposed by General Lon Nol, army chief of staff. Exiled to Beijing, Sihanouk put out a call for support, invited the Communists into his government-in-exile, and created the National United Front of Kampuchea. The National Front included the Khmer Rouge, whose guerrillas rapidly gained control on the ground. Lon Nol benefited from support of the United States, which in 1969 began massive secret air raids that increased in intensity until they were stopped in 1973.

## 1975-1979: THE RISE OF DEMOCRATIC KAMPUCHEA

On April 17, 1975, the Khmer Rouge entered Phnom Penh. The Communist Party ruled over the new nation of Democratic Kampuchea, led by its secret prime minister, Pol Pot, and the country became isolated from the rest of the world. On December 31, 1977, after two

years of increasingly rancorous border incidents, the Communist regimes in Cambodia and Vietnam broke off diplomatic relations with
each other. Cambodia was supported by China, while Vietnam was
backed by the Soviet Union.

## 1979–1998: VIETNAMESE OCCUPATION AND THE ONGOING CIVIL WAR

On January 7, 1979, Vietnam launched a massive offensive and invaded Phnom Penh. The Khmer Rouge regime fell and its armed
forces withdrew to the mountains in the northern and western parts
of the country. Cambodia was now under Vietnamese occupation,
and its leaders were mostly ex–Khmer Rouge cadres who had deserted
the movement and escaped to Vietnam. They include Hun Sen, who
became prime minister in 1985, at the age of thirty-four. At the same
time, the world discovered the magnitude of the crimes committed
under Pol Pot: about one-quarter of the Cambodian population of
Democratic Kampuchea perished. Yet the context of the Cold War
meant that the international community condemned Vietnam's inter-
vention and, for another decade, the Khmer Rouge remained Cambodia's legitimate representative in the United Nations. On September
26, 1989, after ten years of occupation, the last remaining Vietnamese
troops left Cambodia. On October 23, 1991, peace agreements were
signed in Paris. Between March 1992 and November 1993, Cambodia was put under the temporary authority of the UN, which organized general elections. Sihanouk returned to his country as king of
Cambodia again until he abdicated in favor of his son, Sihamoni, in
2004. A coalition government was established between Hun Sen and
the head of the Royalist Party. The Khmer Rouge soon renounced the
peace agreement and carried on as a guerrilla force, but continued to
suffer more and more defections, including that of Brother Number
Three, Ieng Sary, in 1996. In July 1997, Hun Sen ousted the Royalist
prime minister in a coup. His Cambodian People's Party, which had
been in power since 1979, reaffirmed its political dominance.

On April 15, 1998, Pol Pot died in Anlong Veng, the last bastion
of the Khmer Rouge. Brother Number One had been pushed out of

the movement's leadership less than a year earlier. In December 1998, Khieu Samphan, former president of Democratic Kampuchea, and Nuon Chea, Brother Number Two, joined Hun Sen's government, which welcomed them publicly in the name of national reconciliation. Three months later, Ta Mok, the last Khmer Rouge leader not to have surrendered, was arrested. The Cambodian conflict that started some thirty years earlier came to an end.

## 1999-2013: THE POSTWAR ERA

On June 6, 2003, after six years of tense negotiations, an agreement was signed between the Cambodian government and the United Nations to establish a tribunal charged with prosecuting the most senior Khmer Rouge leaders who were still alive. In July 2006, Cambodian and foreign judges chosen to sit on the Extraordinary Chambers in the Courts of Cambodia were sworn in in Phnom Penh. A few days later, Ta Mok, imprisoned since 1999, died at the age of eighty-one. On July 31, 2007, Duch was charged with crimes against humanity and war crimes. Four other senior leaders—Nuon Chea; Khieu Samphan; Ieng Sary; and Ieng Sary's wife, Ieng Thirith, former minister of social affairs—were arrested and indicted at the end of 2007. Duch's trial began in early 2009. He was sentenced to life imprisonment by the chamber of appeals in February 2012. The trial of the four other defendants started in November 2011. Ieng Thirith, eighty years old, was found unfit to stand trial and was released in September 2012. Her husband, Ieng Sary, died in March 2013, at the age of eighty-seven. A judgment of the two remaining accused, Nuon Chea, eighty-seven, and Khieu Samphan, eighty-two, is expected in early 2014.

# ACKNOWLEDGMENTS

To my friends Heather Ryan, constant companion during the trial, and Stéphanie Gée, unflagging source of support during the writing process.

To the friends, old and new, working at the Khmer Rouge Tribunal, who went out of their way to help me.

To those who made my stay and my work in Cambodia so comfortable, Mrs. Kit Kim Huon, Khem Somalay, Lily Luu, David Harding.

To Phann Ana.

To Lin Zi.

Thank you to Antoine Audouard for following my work with patience and wit, to Susanna Lea for having carried out the final stages so impeccably, and to Yvon Girard for his enthusiasm and confidence.

At Ecco, my most sincere thanks to Dan Halpern, Hilary Redmon, and Emma Janaskie, for their great support and work.

In memory of Mike Fowler.

# NOTE ON SOURCES

The author has attended the entirety of the public trial, as well as pretrial hearings, between 2007 and 2010. All quotes or documents referred to in this book are from court proceedings, unless otherwise specified.

Three languages were used during the trial before the Extraordinary Chambers within the Courts of Cambodia (ECCC): Khmer, English, and French. When the speaker was speaking in Khmer, simultaneous interpretation was provided in English (by non-native English speakers), then into French (by native French speakers). Problems of interpretation and translation are common to all international tribunals. There is no reason to believe that, over months or years of proceedings, such problems affect significantly the understanding and the outcome of the case. However, being the product of live interpretation and subject to copy editing, official transcripts should be taken with caution.

In this book, all quotes from English speakers are original. Quotes from French speakers have been retranslated into English from the original French. (By nature, transcripts based on simultaneous interpretation are of limited accuracy.)

The main difficulty was to deal with quotes from Khmer-speaking individuals, including the accused. It was practically impossible to go back to the original speech and have it translated again, as was done from French to English. Yet there was often a need to edit and rewrite in proper English for the sake of clarity. The same was true for some documents from the case file. An example of how official transcripts can depart from what was actually said in court is the preamble of the Constitution of the Kingdom of Cambodia. The official English transcript does not follow the shortened version used in court by the French lawyer; instead, it reads like a mix of official translation and free interpretation. We did not rely on it, and stuck to a strict translation of what the French lawyer chose to read in court. Similarly, some English translations of Khmer Rouge-era documents by the court language services, such as extracts from the Party's newspaper or the Party's statutes, were rather poor, and have been edited accordingly for the sake of clarity, while retaining the somewhat awkward Maoist parlance.